This book is dedicated to

Rosie Héloise O'Connor

A shepherd measuring time at night
From *The Kalendar & Compost of Shepherds*,
first published in Paris 1493.

UNDERSTANDING

TIME

Jolyon Trimingham

IMAGIER PUBLISHING
BRISTOL 2016

First published in 2016
by: Imagier Publishing
Bristol BS35 3SY
United Kingdom
Email: ip@imagier.co.uk
www.imagier.co.uk

Cover and text design by Allan Armstrong.
The clock face on the cover is derived from the
14th century astronomical clock located
in the north transept of Wells Cathedral

Frontis: 'A shepherd measuring time at night'
From *The Kalendar & Compost of Shepherds,*
first published by Guy Marchant in Paris 1493.

The paper used in this publication is from a
sustainable source and is elemental chlorine free.

Printed and bound by Bookfactory.co.uk

Contents

Transcendental Idealism. The primacy of internal time of which we can become aware independent of and antecedent to knowledge acquired through visual perception.

Chapter 2 Experience

Time as experienced for each person is a unique and unrepeatable experience. It cannot be reduced to universal scientific laws or to a fixed algorithmic function. This is a lived time rather than an ex-post facto conceptualised time. On the most general level our experience of time is influenced by our membership of the human species. There are variations of spatio-temporal apprehension according to the number of dimensions a species can apprehend. Discussion of the less universal effects of specific natural and cultural environments on time are then detailed. In particular the effect of the rise to global dominance of our Western culture and technological achievements is considered. This is placed within the context of other global civilisations, in particular Eastern ones, to emphasise that modern assumptions about what time is can be viewed, not just in relation to their 'truth', but also to their historical significance. The chapter closes with a discussion of the individual experience of time. Our mind and our body can be more than just passive recipients of temporal input; it is possible to take positive steps to develop and enrich our awareness of time. Artistic expression or appreciation is one avenue and is examined in the following chapters.

Chapter 3 Freedom

Freedom is not an absolute but, as Henri Bergson maintains, admits of degrees. It varies individually relative to how independent someone becomes of their environment and social conditioning. In expressing this freedom the causal chain of events stretching back through time is loosened. In moments of heightened awareness or concentration the link may actually be broken. This gives rise to a feeling of pure freedom which can be experienced internally as a state of 'timeless' bliss or manifested externally in creative activities.

Chapter 4 Creation 75

Art is communication of the creative impulses and can express the divine which permeates our material existence, but which is elusive and difficult to portray except through metaphors of artistic creation This chapter illustrates how time is felt and presented in artistic expression. Particular attention is given to two art forms. Firstly, music because it is the art form most removed from the domination of our spatial interpretations of time. The second is cinema which uses the already temporal deception of the 'moving picture' and then adds visual techniques to create what Gilles Deleuze describes as 'virtual time'. After this discussion of time in art the next chapter examines time in science before a synthesis is made to outline a new paradigm for time.

Chapter 5 The New Physics and Objective knowledge 94

Many reasons why science alone can never explain time. Time's flow can be reconciled with theoretical science and approximated in applied science as most analogue processes can be handled digitally. But time's direction, its 'arrow' is more problematic. Classical science's treatment of events as theoretically reversible in theory has proved unsustainable. Directionality gradually reinstated. Examples of this focusing on entropy and evolution. Time and Einstein's Theories of Relativity. Consideration of the post-Einsteinian scientific treatment of time starting with Chaos theory. Ilya Prigogine's work on irreversible chemical processes. Chaos theory can just about fit within classical science as modified by relativity but this is not so for quantum theory. Efforts by some quantum theorists to keep both consciousness and a dangerously subjective directional time from threatening to subvert the pure objectivity of physics have not been overwhelmingly accepted. While these 'materialist' quantum physicists speculate on theories of multiple spatial dimensions others lean towards a more 'idealist' approach. Roger Penrose is a moderate version of this with his belief in Platonic Ideas; a more extreme case is Amit Goswami, a self-styled monistic idealist. Quantum theory must admit time asymmetry together with some kind of active participation for human consciousness. Current attempts to construct a neurologically plausible quantum theory of the brain to explain what consciousness is may resolve this debate. With Quantum Theory the barriers separating science from other fields are being taken down. The emergence of revised notions of time by which scientific approaches can be integrated with non-scientific perspectives follow in the remainder of the book.

Chapter 6 What is Time?   140

The threads of the preceding chapters are gathered up and woven together to design a tapestry portraying the conception and perception of time. Time and consciousness as inseparable but nonetheless distinguishable. Spurious distinction between objective and subjective time: they are just different perspectives on the same phenomenon. Non-locality extended beyond usage in quantum theory. Limitations of language to describe the essence of time. Moving beyond words as symbols to a more metaphorical language to describe what their symbolism tries to represent. To make headway this must be combined with non-verbal modes of expression and understanding.

Chapter 7 Metatime

How heightened mental, emotional or bodily sensitivity sharpens our apprehension of time and all that exists within it. Contemplation, love and physical action as paths of development which respectively correspond to those three types of sensitivity. Contrast between external knowledge developed by learning and internal knowledge acquired through concentration. Both of these forms of knowledge augment understanding, wisdom and well-being. Poetic expressions of being in Metatime.

Chapter 8 Conclusion

Understanding and management of external time is arguably essential, particularly in modern technocratic life. But this definitely doesn't need to be at the expense of our intuitive feeling of inner time. In young children this aspect of time predominates over the more practical external time sense which tends to take over in maturity. Any complete theory of time cannot be limited to spatial perceptions and scientific conceptions of it. Room must be set asidet for an awareness of time from all the other equally valid perspectives described in this book. If these are left out then so is the full meaning and beauty of our being alive in time.

About This Book

The book has two purposes, a negative one and a positive one, which are complementary. The negative one is to attack the 'hijacking' of time in popular versions of modern mathematical and scientific thought, which have tended to push out of the picture views of time which do not 'fit in' with modern physics. These share the assumption that discussion of the nature and meaning of time is only possible through knowledge of twentieth century developments in mathematics and physics. Any other approaches to time are ignored. By this exclusion our understanding of the role of time in our lives is impoverished. Despite their attempted monopolisation of the discussion of time their approach has not resolved the paradoxes or thrown light on the mystery of time. The significance of time is too important to be left to physicists and mathematicians alone.

The second purpose is to show how, by supplementing, or sometimes overriding, the scientific perspective, a wider understanding is reached. A purely scientific description is not wrong but incomplete. Time cannot be described in terms of its measurement alone. Such a partial description is like defining a person by his or her physical appearance without any reference to their character. In a less restricted environment time can be understood more fully and paradoxes about time can start to be resolved. Also, a more open attitude towards time can enrich our lives, increasing our motivation and sense of fulfilment. If the scientific explanation is taken as the *whole* explanation it lies by omission.

These two negative and positive aims fuse in the later chapters which demonstrate how the scientific use of time can be reconciled with non-scientific views on time. In the light of this a new insight into the nature of time is explored. This book considers time from a number of perspectives, such as time in art and music, time and love, time and history, time and death, as well as looking in detail at how time is treated psychologically in everyday life. Profound insights on time can be found in non-scientific approaches which focus on aspects of time, arguably the more important aspects, which are either neglected or belittled in science. A glance at the references will reveal some of these sources. The scientific approach to time is only one approach, and not necessarily the right approach to the moral, intellectual or practical concerns which are related to our understanding of time.

Both the terms 'New Physics' and 'Post-Science' are slippery, so to try to avoid misconceptions I indicate at the outset my understanding of them.

The term New Physics has been left almost deliberately undefined by scientists who resent the innuendo that in some way physics is not a continual progress of knowledge. From the perspective of debates about time, relativity can indeed be seen as a progression, rather than a break with the past. Although 'Newtonian' time is shaken, time in relativity theory is still independent of our consciousness (though not of 'an observer'), has no directionality and is regarded as definable in terms of its external measurement. But for the other main spur of the new physics, quantum theory, this is not the case. It is not a development from the tenets of classical physics but, in many ways, a new departure, particularly in regard to the role of time.

The effect on this book is that in discussion of the new physics, attention will be focused on quantum theory rather than relativity. Quantum theory discards both classical scientific and common sense notions, but it works. A common misconception is that it applies only to the incredibly small, so that it can safely be ignored in 'everyday life'. This is not true. Two examples are laser beams and superconductivity which exist at the macro level and can only be 'explained' by quantum theory.

Quantum theory opens the door to 'post-scientific' thought in finally breaking down the classical Cartesian divide between consciousness and the material world. Any absolute distinction between an objective world and an interior world is in doubt. Eminent scientists now feel they are entitled to discuss the subjects which were previously outside their remit, as being non-objective. But this means they are moving into areas where other non-scientific views also need to be taken into account. Science is escaping from its cage, modifying and being modified in its turn, within a wider paradigm. This is why for some we are at the beginning of what has been dubbed a 'post-scientific age'.

I wish there was a better term, as 'post-science' could imply:

1. that science is being superseded rather than supplemented
2. that 'post-scientific' philosophy is based on new insights.This is not the case. Its main emphasis is on the need for influences from different disciplines, old or even ancient, as well as recent, to counterbalance what it views as 'scientific chauvinism'.

For these reasons, the term will be avoided in the text. Nevertheless it is a useful shorthand for the current reaction against over-reliance on the scientific approach.

The term 'post-science' can be used in a 'weak' sense. It does not indicate that science is 'wrong', only that science is incomplete when taken as the *whole* explanation of nature and life. "A moderately post-scientific age is one that still relies on science yet is more conscious of its limitations and sceptical about its potential to adequately address all of life's demands and provide all of its answers" (Lovat: 2003).

'Strong' post-scientific philosophy on the other hand has an altogether different perspective. Allying itself with post-modernism it views all Western science as a cultural 'metanarrative', a product of Western history and thought modes, with no claim to be considered as revealing any objective truths. Nevertheless, the material and medical benefits of the West are acknowledged. So it is left in the strange position of accepting the benefits whilst rejecting the methodology and concepts which were a prerequisite to achieving them.

This is certainly not the view taken here. In contrast to any post-modernist relativism the view taken is that there are insights, particularly about the nature of time, which although refracted through different cultural filters, are essentially different manifestations of the same truth. This does not privilege Western thought – the 'New' Science in particular, but leaves space for an accommodation. Hopefully, it will not be just an accommodation, but a cross-fertilisation with the New Physics occupying the stage, but not the centre stage in an integrated description of time. This would include time in both its subjective and objective manifestations and thus help resolve the paradoxes which crop up if only one or the other is believed to represent the 'truth'.

About This Book

A note on quotes:
Normally when quotes are cited they are cross-referenced to the bibliography. However, when a work is quoted which is not included in the bibliography the source has been placed in the text itself, e.g. "Stay Stay moment thour't so fair" (Goethe, Faust Part II).

Introduction

"...the physicists' model of time has been contrasted with that of our personal experience, full as it is with weird psychological images and paradoxical motion. The grey area between mind and matter, philosophy and physics, psychology and the objective world is only the threshold of exploration, yet any ultimate picture of reality cannot omit it. It could be that the images of time so dear to us – the existence of the present moment, the passage of time, free will and the non-existence of the future, the use of tenses in our language – will come to be regarded as only primitive superstitions that spring from an inadequate understanding of the physical world......on the other hand, it could be that for *once our minds are more reliable than laboratory instruments, and time really does have a richer structure than we perceive.* In which case the nature of reality, of time, space, mind and matter will suffer a revolution of unprecedented profundity." (Davies: 1980,199 my italics)

In recent years a great number of books have been written by scientists and scientific popularisers about time and the nature of time. These are often dull and tedious. Three examples out of many are (1) Le Poidevin, Robin (2009) *Images of Time: An Essay on Temporal Representation*), Oxford University Press, (2) Dennet, Daniel (1996) *Darwin's Dangerous Idea: Evolution and the Meanings of Life*, Penguin Science Paperback or (3) Mellor, D.H. (1998) *Real Time II*, Routledge. People buy these books in the mistaken hope that these works will throw light on the mysteries of the nature of life, explain the origins of the universe and give more meaning to our transitory existence in time. Instead of providing illuminating insights the books tend to explain how time is really understandable only through arcane concepts of relativity and quantum physics. They then proceed to describe these in a condescending or misleadingly simplistic fashion. Many intelligent and talented people, after failing to fully grasp the scientific concepts 'explained' in these works, may begin to doubt the validity of their own experience, understanding and opinions about time. But those looking for enlightenment on what time is all about and how it affects our lives by reading books 'popularising' physics, neurology, mathematics, cosmology or astronomy are not looking in the right place. These books may be by respected experts in their disciplines but, without intending any disrespect to them, they cannot be allowed to have an unchallenged monopoly on the subject of time.

The purpose of the scientific approach to time is usually to use the analysis for functional purposes to further scientific progress and not to answer basic human curiosity about why and how we exist in time and its effect on feelings and actions. There is nothing wrong with this in itself. What is wrong is the attitude that this is the only approach which is 'objectively' valid. Other approaches are either ignored ('scientism') or perceived as relevant only insofar as they can be described in scientific terms ('reductionism').

Brian Appleyard, writing about his meeting with Stephen Hawking was dismayed that Hawking believed that modern physics "invalidated other disciplines". He also believes that Hawking's 'Brief History of Time' "was so successful that it resulted in a massive wave of popular science books, all of them sold to publishers and readers on the basis of their wild claims for the competence of contemporary science and almost none of them displaying a trace of humility" (Appleyard: 2002). For Appleyard, Hawking is "an emblem of scientism...

the figurehead of a movement that I believe has promoted contempt for human, and humane wisdom" (Appleyard: 2002). Hawking's follow-up book with an equally arrogant title "The Universe in a Nutshell" shows his ignorance of, and contempt for, other disciplines apart from physics. In an interview in the "Daily Telegraph" he talks about the 'positivist philosophy of science which I adopt" (Hawking: 2001), apparently unaware that most scientists abandoned positivist philosophy many years ago. In the book itself he talks about a "few setbacks (in the advance of knowledge), like the Dark Ages after the fall of the Roman Empire" (Hawking: 2001, 157) which betrays an ignorance of the history of his own discipline, mathematics. It was after all, during these so-called Dark Ages that the Arabs created algebra.

Human beings are seen solely as computers. "The brain does not have a single CPU – central processing unit – that processes each command as a sequence. Rather it has millions of processors working together at the same time. Such massively parallel processing will be the future for electronic intelligence as well" (Hawking: 2001, 169). We need to get our act together before computers overtake us:

"In a way, the human race needs to improve its mental and physical qualities if it is to deal with the increasingly complex world around it and meet new challenges such as space travel. Humans also need to increase their complexity if biological systems are to keep ahead of electronic ones. At the moment computers have the advantage of speed, but they show no sign of intelligence...But computers obey what is known as Moore's law...it (computer growth by Moore's law) will probably continue until computers have a complexity similar to that of the human brain." (Hawking: 2001, 165) His conclusion from this is that genetic engineering is essential to increase the size of the brain. The big problem, for Hawking, is that the brain will become too big to pass down the birth canal. "Having watched my three children being born, I know how difficult is for the head to get out. But within the next 100 years, I expect we will be able to grow babies outside the human body, so this limitation will be removed"(Hawking: 2001, 167). So, problem solved. At this point in reading this work by the Lucasian Professor of Mathematics at Cambridge, I don't know whether to laugh or cry.

The general belief in the reduction of all human knowledge to something that can be coded is a widespread article of faith. Any information that cannot be coded, such as the flow and direction of time, is condemned as subjective and therefore not real. This is just nonsense. You reading this book, appreciating a good joke, a shared memory, feelings of love, the atmosphere at a concert all exist without being quantifiable. This ultimate folly contends that all phenomena, mental or physical, can be expressed in theory in computational terms. Stephen Wolfram in his huge book published in 2002 takes this to the extreme with his theory of computational equivalence (Wolfram: 2001).

If, from the point of view of our understanding we are rather slow computers, from the point of view of our actions we are sophisticated robots. "What on earth do you think, you are, if not a robot, albeit a very complicated one?' asks Richard Dawkins (Dawkins: 1976, 270). Unlike, the computer analogy the robot analogy is not often used as it is simply taken for granted, ever since Laplace formulated very clearly the universal determinism that lies behind it.

"Let us imagine an Intelligence who would know at a given instant of time all forces acting in nature and the position of all things of which the world consists; let us assume, further, that this Intelligence would be capable of subjecting these data to mathematical analysis. Then it could derive a result that would embrace in one and the same formula the motion of the largest bodies in the universe and of the lightest atoms. Nothing would be uncertain for this Intelligence. The past and the future would be present to its eyes" (Pierre-Simon Laplace, Theory of Probability, 1812). The analogies of robots and computers support each other in that if we are robots all of our thoughts and actions can be viewed as an analysable sequence and as analysable sequences they can be reduced to binary arithmetic for computer data entry.

Although the above has focused on Stephen Hawking's view, his dogmatic approach is shared by many scientists and philosophers of science, who feel threatened by any approach that is not reducible to quantifiable data and therefore mathematically expressible. Another prominent outspoken example is the Australian philosopher of science, who specialises in space and time, J.J.C. Smart. He is a follower of 'physicalism' with its 'identity;' theory of mind, holding that states and processes of the mind are identical to states and processes of the brain, and therefore quantifiable. There is therefore no such thing as an internal 'mind' time – it is an illusion.

As we shall see later the materialist dogma has to be modified even in physics when quantum theory is considered. "Perhaps the most important, and the most insidious, assumption that we absorb in our childhoods is that of the material world of objects existing out there – independent of subjects, who are observers.......whenever we look at the moon, for example, we find the moon where we expect it along its classically calculated trajectory. Naturally we project that the moon is always there in space-time, even when we are not looking. Quantum physics says no. When we are not looking, the moon's possibility wave spreads, albeit by a minuscule amount. When we look, the wave collapses instantly; thus a wave could not be in space-time. It makes more sense to adapt an idealist metaphysic assumption: There is no object in space-time without a conscious subject looking at it" (Goswami: 1993, 59-60)

This introduction is not meant as an attack on science. It is just to put it in a wider perspective to counter what is occasionally an overweening arrogance. The tremendous power of science and the stupendous technological forces it has engendered in its application should not blind us to other parts of life where its achievements have been minimal. Scientific understanding has focused almost exclusively on helping us to manipulate matter. This ranges from the very small as met with in atomic physics and quantum mechanics, to our own bodies in medicine through to the inconceivably large as we endeavour to comprehend the universe.

Consciousness which gave rise to these insights, however, and everything associated with it, still resists scientific progress. According to some scientific pundits this is because such 'sciences' as psychology are still in their infancy. Whether this be true or not, a terrifying imbalance has built up between our mastery of the world and our knowledge of ourselves

An exclusive narrowed-down single focus is dangerous. There is something soulless and empty (and basically untrue) about a purely objective approach. When scientists stray beyond their job as interpreters of the physical world and attempt to lay down truths concerning the nature of time or consciousness they are straying beyond objective analysis and need to be reminded that on these levels their insights are no more valid than those arrived at from other disciplines and that the objectivity of the 'scientific method' is not the only true measure.

Natural counterweights of *wisdom, love* and *self-knowledge* are ignored: the three quotations which follow are examples which stress the importance of moral, emotional and cognitive insights, which would be meaningless if all knowledge was based on scientific analysis.

Wisdom: "One who has not abstained from evil conduct, who is not tranquil and Self-controlled, whose mind is not at rest, cannot attain Atman by mere book-knowledge or mundane cleverness" Katha Upanishad (II-24).

Love: "If I speak in the tongues of men or of angels, but do not have love, I am only a resounding gong or a clanging cymbal. If I have the gift of prophecy and can fathom all mysteries and all knowledge, and if I have a faith that can move mountains, but do not have love, I am nothing." (St. Paul 1 Corinthians 13: 1-2) Self-knowledge: the inscription at the Temple of Delphi sums it up "Know Thyself and Moderation in all things"

Subjectivity is as valid and sometimes as universal as objectivity. It also has precedence in that there could be no objective knowledge without conscious subjects to make the distinction. A number of scientists, especially in areas that are connected with investigating mental activity, have appreciated this. Some of these, for example Antonio Damasio, prominent neurophysiologist and author, will be referred to later.

The primacy of the subjective applies especially to an understanding of time. The flow, directionality and indivisibility of time require a conscious subject. The whole history of Western science (before the twentieth century) can be viewed as an historical aberration demonstrating what happens if time is totally objectified. As a working hypothesis this may have allowed the phenomenal progress in learning how to understand the objective world, but at the same time it may account for the lack of progress in understanding the subjective world of consciousness.

A more moderate and modest scientific approach is that of Paul Davies, who has written much on the nature of time from the perspective of a physicist. Nevertheless, despite his obvious mastery of the subject and ability to transmit this both to the expert and the layman, he seems unable to view the nature of time from outside the unproved assumptions of the new physics. Reading his work sometimes between the lines, sometimes in his explicit statements, he reveals the Western approach to time with both its powers and the weaknesses. The following assumptions are made.

1. He accepts without serious discussion that there are no valid views of time and its nature outside the Western scientific tradition from Newton through to Einstein and Quantum Theory. "I shall deal with physical measurable time as if it is real. For that is the founding assumption of science – that there is a real world out there that we can make sense of. *And that world includes time.*" (1995: Davies, my italics). The remaining mysteries of time such as its flow or 'arrow' are only there because the scientific analysis of 'subjective' time which has these elements lags behind our knowledge of 'objective' time. "Attempts to explain the flow of time using physics, rather than trying to define it away using philosophy, are probably the most exciting contemporary developments in the study of time. Elucidating the mysterious flux would, more than anything else, help unravel the deepest of scientific enigmas – the nature of the human self.....in my opinion, the greatest outstanding riddle concerns the glaring mismatch between physical time and

subjective, or psychological time.(1995 Davies, 278-283) " He sees only two solutions. " Does it (psychological time) reflect an objectively real quantity of 'time' out there in the world of material objects that we have simply overlooked? Or will the flow of time be proved to be an entirely mental construct – an illusion or a confusion after all". (1995: Davies, 283)

2. His methodology rests entirely on the computability of all knowledge. "All information about the world can in principle be represented in the form of binary arithmetic (ones and zeros) this being the form most convenient for computer processing" (1993: Davies, 80.). It follows that "time (for the physicist at least) is that which is measured by clocks" (Davies: 1995, 58)

He takes for granted the assumption of a linear progress of human knowledge and capacities as scientists come closer and closer to the 'truth', which will be looked at it in much more detail later in this book. "The association of time with the mystical and the organic, fascinating and compelling though it may be, undoubtedly served to hinder a proper scientific study of time for many centuries" (Davies: 1995, 29).

The 'problem' of the 'arrow' of time is addressed not in terms of the meaning of an individual's lifespan and mortality but rather the meaning or paradox it poses for scientists, especially cosmologists and astronomers. "Most scientists are agreed that the source of the asymmetry – time's directionality – can ultimately be traced to cosmology and the large-scale behaviour of the universe, but the precise nature of the connection remains obscure and contentious. " (Davies 1995, 282).

Davies has a superb grasp of time and for the understanding of its significance in modern science and an account of how this has developed there is probably no better source. There is no intention here to dismiss or underestimate the success of relativity and quantum physics in our understanding of the world nor their achievements when applied to technology. Nor is the understanding of time on which they are based contested. However, the approach here is iconoclastic in that it is designed to destroy the image of time presented by them as the only time which is real, by placing it in a much wider context. Davies, in particular in his book 'About Time', presents the scientific view of time with all its current conundrums so well that it seems to mirror the truth. In fact, it probably is the truth. The fundamental objection is simply that it is not the whole truth. The scientific story is correct but it is not the whole story and without the whole story there is the old fallacy of mistaking a part for the whole. Of course, Paul Davies is a distinguished and dedicated scientist, so it is not surprising that he adheres to the scientific understanding of truth. Nevertheless, there is a hankering after a wider perspective which glimmers through his writings, which are full of references to non-scientific views of time from the Australian aborigine dream time to Plotinus's mysticism, which are also mentioned in this book. His comments about an encounter with the Dalai Llama reveal his ambivalence:

"In a lecture in London some years ago, I found myself sharing the platform improbably with the Dalai Llama. Our task was to compare and contrast time as it enters into Western scientific thinking and Eastern philosophy. The Lama spoke with quiet assurance, but unfortunately in Tibetan. Though I tried to follow the translation for enlightenment, I didn't receive much. Culture clash, I suppose.

After my lecture, we had a tea break, and the Dalai Lama took my hand as we walked out of the building into the sunshine.....I had the overwhelming impression of a gentle and intelligent man with insights of value to us all....I came away from the occasion with a deep sense of missed opportunity". (Davies: 1995, 22).

Davies actually states the objections (enlarged upon in this book) of the non-scientist to the scientifically dominant view but then draws back:

"In appropriating time for themselves, and abstracting into a stark mathematical parameter, physicists have robbed it of much of its original human content. The physicist will usually say 'Ours is the *real* time, all that there really is. The richness of human psychological time derives entirely from subjective factors and is unrelated to the intrinsic qualities of real physical time' – and then goes about his or her work and daily life immersed in the complexities of human time like everyone else." (Ibid, 273).

Should we simply shrug the human experience of time aside as a matter solely for psychologists? Does the time of consciousness, particularly in heightened or altered states, have no relevance at all to the time of Newton or Einstein? Does our impression of the flow of time, or the division of time into past, present and future, tell us nothing about how time *is*, as opposed to how it appears to us muddle-headed humans?

"As a physicist, I am well aware how much intuition can lead us astray.....Yet, as a human being, I find it impossible to relinquish the sensation of a flowing time and a moving present moment. It is something so basic to my experience that I am repelled by the claim that it is only an illusion of misperception. It seems to me that there is an aspect of time of great significance that we have so far overlooked in our description of the physical universe" (Ibid, 275). But after the stimulating insights which follow a few pages further it is clear that the solution to this is not going to be outside of physics. "Attempts to explain the flow of time using physics, rather than trying to define it away using philosophy, are probably the most exciting contemporary developments in the study of time. Elucidating the mysterious flux would, more than anything else, help unravel the deepest of all scientific enigmas – the nature of the human self" (Ibid, 278).

He is not quite prepared to cross the bridge over from the scientific view to associate it with the arts, philosophy and mysticism as others of a basically scientific background have done. Maybe that is his strength as it results in a cogent and convincing exposition of the nature and benefits of the mainstream scientific view.

Nevertheless, the belief that science has a monopoly on understanding of time as in the above examples, is not shared by all scientists. These include Niels Bohr and best-selling authors such as Fritjof Capra or Julian Barbour who have a wide enough vision to step outside their discipline. But such contributions are still the exception. Reading them can give the impression that criticising the effects of the scientific community on our views of how the world is in time is now like chasing a paper tiger.

A striking and eerie topical example that mainstream science does not have a special claim to the truth (and of the convergence of recent scientific thought with other traditions) concerns the 'Big Bang' which created the cosmos. The theory put forward by Neil Turok of Cambridge University and Paul Steinhardt of Princeton in 2002 is of a cyclic universe. The big bang was not a one-off event but is a repeatable phenomenon. When after trillions of years the universe runs down the energy that fills new radiation and matter are created starting a new cycle. The parallels between their hypothesis, which is being taken very seriously by the scientific community, and ancient Vedic cosmology are quite stunning. Modern cosmology is rediscovering the cosmic pattern of expansion and contraction, creation and destruction symbolised by Indian goddess Shiva over the same awesome timescales used by the Vedic astronomers over two thousand years ago.

If readers took reductive scientific views at face value it would strip them of the confidence and trust in their own judgement and abilities. It would be an arid life if the following three examples of how the world is seen came to be threatened by an unquestioning acceptance of these views.

1 Our common sense understanding of time.

This is to be rejected because we don't understand enough. Only experts can understand the real nature of the world and the role of time. Our everyday understanding does not square with the scientific view. In objective time there are no happy times, sad times, boring times, sleeping times or waking times. Subjectively, we don't measure our time as a commodity, but from the outset feel it qualitatively. Talking to a father pushing his daughter back and forth on the swing in the local park he refers to this time as 'quality time'. At the other end of the scale, I have spent months of my life in deep depression unable to do much at all or to enjoy the little I did do. Yet I don't feel that I have to make up the loss in some kind of quantifiable way. There is a tendency to belittle this internal time as subjective, and therefore untrustworthy and 'unverifiable'. But although parts of our internal time are purely relative to an individual much of the internal sense of time applies universally to human beings. Indeed, without this no experience at all would be possible for human beings. The flow and arrow of time, the feeling of the present, transience: none of these are part of objective time. *From now on the notion of 'subjective' time with the implication that it is not 'valid' will be dropped and replaced by the notion of 'internal' time.* Just because internal time cannot be objectively measured and varies from individual to individual, this does not make it less real than 'objective' time. In fact, it shall be argued it is more real.

2 Faith in our ability to act meaningfully in the world

This is weakened because all acts (except those on the micro level) are predetermined. We are therefore impotent but suffer from the illusion that our acts are voluntary. What can we or should we do with our lives? All this is bound up with our existence in time. We live in time and die in time. Is our life just "a gilded execution chamber with nature replenished continually with new victims" (Dunne: 1934). This feeling of a pointless conveyor belt towards death is the inescapable conclusion from a narrowly reductionist perspective. No wonder they cannot come to terms with the arrow of time. Paul Davies quotes Professor Peter Atkins, an associate of the British National Secular Society, expressing these 'sentiments': "All change, and time's arrow, point in the direction of corruption. The experience of time is the gearing of the electrochemical processes in our brains to this purposeless shift into chaos as we sink into equilibrium and the grave" (Davies: 1995, 196). This is not just the view of a few extremists – it is fast becoming a scientific orthodoxy. The following quotation is from the conclusion to a book popularising science by a widely read author.

"Some illusions are programmed so firmly into our brains that the mere knowledge that they are false does not stop us from seeing them. Free will is one such illusion. We may accept rationally that we are machines but we will continue to feel and act as though the essential part of us is free from mechanistic imperatives."

"The illusion of free well is deeply ingrained precisely because it prevents us from falling into a suicidally fatal state of mind – it is one of the brain's most powerful aids to survival......."(Carter: 1998, 205-6)

Rita Carter spells out very clearly the consequences of her reasoning.

"I also hope that the ability to modulate brains will be used more widely to enhance those mental qualities that give sweetness and meaning to our lives.....Future generations will take for granted that we are programmable machines just as we take for granted the earth is round."(Carter: 1998, 206)

Perhaps we ought not to fret about this, listen to the 'experts', then follow the advice of Ecclesiastes and just get on with it. "Man cannot find out the work that God hath done from beginning even to the end. I know there is nothing better for them than to rejoice, and do good so long as they live, and also that every man should eat and drink and enjoy good in his labour..." (Ecclesiastes, Chapter 7)

But such a fatalistic attitude is not mandatory. Our experience is not a deterministic given. No knowledge of our character or genetic makeup can ever make our thoughts or actions necessarily predictable. Our sense and conception of the passage of events in time can be adjusted, manipulated or suspended: above all it can be developed. This can happen at crucial moments of decision, when, momentarily, the baggage of our habitual reactions is discarded. It can occur involuntarily as well. As anyone knows who has been in a life or death experience, like a car crash, events slow down to allow our thought processes the time to reflect on what to do. But it does not need crisis situations to bring this about. Such a change can be cultivated by intellectual, emotional or physical methods.

Just as the acuteness of our space sense can be developed by using instruments, such as microscopes and telescopes, so our sense of time can be intensified. In this book I shall consider a number of ways this may be possible, different people taking the ways most suited to their skills and personality. Getting beyond 'objective' time, 'transcending' it comes with an awareness of a movement towards a less programmed existence.

3. Appreciation of the mystery and beauty of life

This appreciation is vacuous. From the point of view of the 'theory of everything' it is meaningless. Why are we here? This great mystery is intimately bound up with time in which we are aware of our own mortality. Does part of us somehow transcend our temporally bounded existence during our life or after our death 'Then shall the dust return to the earth as it was: and the spirit shall return unto God who gave it' (Ecclesiastes 12:7). Even if it doesn't, what should we do with our lives to live them fully? We are not just anonymous observers but mindful individuals filled with curiosity and capable of judgements, feelings and responses. All this is abstracted by the scientist but the internal experience of time is as important, and certainly antecedent to the scientists' time. To leave this aspect of time out is like 'knowing the price of everything but the value of nothing'. It strips us of feelings of wonder, marvel or awe – our sense of the numinous.

There are truths here which cannot be expressed logically or analytically as 'A=B' or sequentially as 'if A then B', because they involve consciousness, our temporal existence and mortality. Analysis cannot reach this because it has to break consciousness or life, or time in which they become manifest, into parts, thus losing their fundamental wholeness. For practical purposes we can separate out emotional states of consciousness and the times at which they occur as measured by the clock. To state that for example at 9.25am X felt angry and as a consequence threw a plate at Y at 9.26am sounds a perfectly sound explanation, but such an analysis only works in the most facile cases. Supposing X resisted the urge to throw the plate. We then have the scenario of two emotions contesting for control of X's actions, which creates three 'entities:' a self, an emotion of anger and a balancing effort to restrain the anger. The analysis becomes even more unreal for less superficial events, such as falling in love, writing a novel, being moved by a superb view or an inspiring work of art which moves our whole being, rather than considering conditioned or unthinking responses.

Our minds are not pieces of machinery in which each component serves a certain function in relation to the whole. Moments of conscious life are by their very nature unique and therefore unrepeatable. Meaning or truth are felt differently each time and at different times of life, of the season or of the day and at each moment. "Conceive it as a power of an ever-fresh infinity, a principle unfailing, inexhaustible, at no point giving out, brimming over with its own vitality. If you look to some definite spot and seek to fasten on some definite thing, you will not find it". This is a description from Plotinus (Enneads, VI, 5:12) of the noetic world, of pure apprehension in the mind, which is not reducible to binary form.

Chapter One

Philosophy

"In the arithmetic sequence, for example the 3 is before the 4... yet the 3 is not earlier than the 4 on this account. Numbers are not earlier or later because they are not in time at all...Once time has been defined as clock time then there is no hope of arriving at its original meaning again" (Heidegger: 1992, 18)

What is time? The usual response when this question is posed in books or articles about time is to quote St Augustine's from *Confessions*. The following quotation appears in most commentaries on time, and this one will not be an exception. "Si nemo me quadrat, scio, si quaerenti explicare velim, nescio" ("If no one asks me I know; if someone asks me to give an account, I don't know".) In other words, the problems start when we try to define it.

Much philosophical discussion of time has centred around trying to ascertain whether the notion of time has its origin in the mind, or whether it is a property of the world and therefore independent of the mind. These points of view are given below as thesis and antithesis.

Thesis: Time is in the mind

Time cannot be perceived through any of the senses. It has no visibility, taste, touch, sound or smell. Take any object in the external world, a chair, a perfume, another person. Remove them or remove the external world in its entirety, and time is still with us. "Time is merely a subjective condition of our intuition... independently of the mind of the subject it is nothing" (Kant: 1934: 47)

Antithesis: Time is outside the mind

Time is independent – the world must exist in time regardless of whether any human or other sentient being is around to be aware of it. Therefore, even if all conscious life disappeared there would still be time. Hence the famous Newtonian definition about time flowing "equably without relation to anything external".

Two seemingly equally valid, but apparently irreconcilably opposed arguments, are what Immanuel Kant called an antinomy, a contradiction between two laws. Our apprehension of time has to be an element in all awareness and but it must also somehow precede that awareness. As everything to be known by us must be known in time, these contrasting definitions of time are part of a greater antinomy: is the world a creation of the mind and therefore 'ideal' or does the world exist independently of mind and is therefore 'real'?

An inquiry limited to the nature of time broadens rapidly into this wider territory of the debate between idealism and realism. To attempt to resolve this here would be presumptuous to say the least: the dispute has gone on for millennia. All that will be attempted now is to highlight the implications it has for a study of time.

If we can show that *everything* is in the mind including the 'outside' world, then there is no longer any antinomy. This is extreme philosophical idealism. Time and everything in time has no existence outside consciousness. The notion of the existence of a world independent of consciousness is an illusion and that includes any objective time. But then the continuous existence of the world becomes a problem. What sustains this creation of consciousness when no one is perceiving it? Bishop Berkeley, when pushed, maintained this was God. Alternatively it could be a kind of universal cosmic consciousness which permeates all.

If we go to the opposite extreme and maintain that *everything*, including time, exists only in the world, then the antinomy again disappears. Time passes without a conscious mind to perceive it and we learn about time, just as we learn about anything else, through experience. This is extreme philosophical *realism*. The problem with this extreme is how an individual consciousness, independent of the world, can arise. Just as extreme idealists are driven to deny the existence of matter, extreme realists attempt to deny the reality of a unified consciousness. For Hume it is the independent continuous existence of a self over time which is problematic. We are nothing more than a "bundle of perceptions" and there is no independent self or "soul":

"It must be some one impression that gives rise to every real idea. But self or person is not any one impression, but that to which our several impressions and ideas are supposed to have a reference. If any impression gives rise to the idea of self, that impression must continue invariably the same, through the whole course of our lives; since self is supposed to exist after that manner. But there is no impression constant and invariable. Pain and pleasure, grief and joy, passions and sensations succeed each other, and never all exist at the same time. It cannot therefore be from any of these impressions, or from any other, that the idea of self is derived; and consequently there is no such idea." (Hume: *Treatise* I: 4: v)

If you have a thesis such as 'time is in the mind' and its antithesis 'time is outside the mind' these could be resolved by a synthesis which unites both the opposing arguments in a larger whole by introducing a new element. Now can a third element be introduced to reconcile the two arguments? There is another phenomenon which, like time, is a condition of any object being known and yet not a property of the objects. This is space.

Without space there can self-evidently be no outside world. Indeed, without space there can be no distance separating the observer and the observed and therefore no 'objective' knowledge. While time alone may be necessary to consciousness, *both* time and space are needed for an awareness of the world, or for any externality, including *objectively perceived time*.

Expressing this in Kantian terms *all* our experience must be intuited and unified in time and all our *external* experience, i.e. objective knowledge, must be intuited in space as well. So, in the case of time there are two modes of intuition. There is an 'internal' time relative to our consciousness as a whole and an 'objective' time, expressible in space, which is relative only to our visual perception. But in order to 'objectify' time in space it must be detached from temporal qualities of consciousness, which cannot be objectified – direction, irreversibility and indivisibility.

Time extended in space is therefore an incomplete time. But the material 'realist' solution to the antinomy is precisely this reduction of time to the spatially extended measurement of time. The view of time from this perspective is that the notion of internal time is totally subjective and therefore an illusion. "The flow of time is unreal, but time is as real as space.... nothing other than a conscious observer registers the flow of time. A clock measures durations between events much as a measuring tape measures distances between places; it does not measure the "speed" with which one moment succeeds another. Therefore it appears that the flow of time is subjective not objective. This illusion cries out for explanation..." (Davies: 2002, 27–8).

But in fact, it is objective time which is the illusion, observable only relative to our visual modes of perception. "Every object of knowledge is known not as a result of its own nature, but of the nature of those who comprehend it" (Boethius: 1969, Book V, VI): this applies to the perception of time.

This observed 'objective' time is not the same as internal time. It is divisible and reversible. It is also not "real" time in that perception is not absolutely instant but follows milliseconds after the event has occurred. Hence we even see ourselves as physical objects in this manner "We always see ourselves from the outside and with the eye of another in retrospect" (Hampshire: 1959, 188). Despite the fact that equating time with its measurement has produced a great number of practical benefits, it is a fundamental error to identify the true 'nature' of time with its functional use as clock time.

If we are going to use space as a kind of *tertium quid* with which to settle the antinomy this must be on the side of mind rather than matter. This is because 'mind-time' is complete and self-contained, whereas objective time only represents time as perceived. Our internal awareness of time is not just relative to our visual observation, but to our sense of self, our other senses such as hearing or touch, our emotions and our understanding. Time, unlike space, does not shut down when eyes are closed.

Is it possible though to have any knowledge of an internal time? Kant maintained that we could not have knowledge of anything without the internal sense of time:

"With regard to phenomena in general, we cannot think away time from them, and represent them to ourselves as out and unconnected with time, but we can quite well represent to ourselves time void of phenomena. Time is therefore given a priori. In it alone all reality of phenomena is possible. These may all be annihilated in thought, but time itself, as the universal condition of their possibility, cannot be so annulled. "
(Kant: 1934, 48).

But though we must acknowledge its 'reality', it cannot be known. According to Kant all perception, on which concepts are based, is conditioned. Therefore it is limited by our spatio-temporal awareness which senses objects as phenomena in *our* time, but not as they are apart from that, i.e. as they are in themselves. So although we must logically presuppose something giving rise to a phenomenon we can never cognize it. He called his system 'Transcendental Idealism'. This applies to time. There must be unconditioned time, though we can never cognize it:

"We deny to time all claim to absolute reality; that is, we deny that it, without having regard to the form of our sensuous intuition, exists absolutely in things as a condition or property. Such properties as belong to objects as things in themselves, can never be presented to us through the medium of the senses. Herein, therefore, consists the transcendental ideality of time, according to which, if we abstract the subjective conditions of sensuous intuition, it is nothing, and cannot be reckoned as subsisting or inhering in objects as things in themselves, independently of its relation to our intuition."
(Kant: 1934, 51)

To remove any trace of dualism, including Kant's transcendentalism, it is necessary to step outside mainstream

Western philosophy. Internal time is only knowable if we admit that there are other ways of knowing besides perception. For instance, Shankara, generally recognised as one of the greatest Indian philosophers, in the late seventh century used the principles of different 'levels' of reality which could be reached by developing of consciousness. In the West, Phenomenology has blurred the distinction between conscious perception and the objects of that perception. Both Shankara and Phenomenology assume the possibility of lifting 'the veil of perception' to allow awareness of a pure time without or prior to its spatial analogue. For this the subject-object, internal-external, mind-matter, knower-known divisions must be collapsed by unifying them in a whole. "The real maya is the separateness. Feeling and thinking that we are *really* separate from the whole is the illusion"(Goswami: 1993 195).

In this non-dualist mindset, objective and internal time are not two different times. Objective time is a reflection in the external world of internal time. To know this internal time we must look within, seeing the world outside as partly what we make of it rather than having any absolute objective independence. Any notion of us as mere passive perceivers for whom the mind is just a *tabula rasa* must be abandoned if our internal time sense is to be apprehended.

Chapter Two

Experience

"You ask me about the idiosyncrasies of philosophers?...There is their lack of historical sense, their hatred of even the idea of becoming, their Egyptianism. They think they are doing a thing *honou*r when they dehistoricise it, *sub specie aeterni* – when they make a mummy of it. All that philosophers have handled for millennia has been conceptual mummies, nothing actual has escaped from their hands" (Friedrich Nietzsche *The Twilight of the Gods*)

The aim of this chapter is to provide a complete contrast with the preceding chapter, looking at time in which we live, breathe and have our being. That is to say the time in which we eat, sleep, dream, work, make love, laugh and cry, rather than the time during which we think or even time in which we theorise about time.

There is a sense of intensity, sometimes of urgency in time as lived, which is absent from the logic of philosophy and natural science. Both these rely on squeezing time dry of essential flavours of different moments of time, by assuming that all moments of time are discrete and homogenous. Events in this desiccated time are repeatable like philosophical arguments in which the same premises result in the same conclusions, or experiments in which this logic is applied, so that the same conditions result in the same results.

All can therefore be expressed or made to occur using what in computer science is called an algorithm, a programme which is a sequence of logical steps each requiring a binary 'yes' or 'no' answer which will lead ineluctably to the same result. This is a very useful resource in information processing, for instance turning our thousands of identical hacksaws or programming the whole process of a company's sales order, registering it, calling on the inventory, dispatching it, invoicing and adjusting the company's accounts. But it is an absurd approach when describing living events or the time in which they occur, the essence of which is that each moment is unique and unrepeatable. Life is not black and white binary. It is much more colourful.

Many aspects of time can be understood by anyone with insight. You don't need to have a degree in computer science, astrophysics or mathematics to understand time. We live in it before we think in it and we think in it before we reflect on what it is.

Unable to use algorithmic logic, conclusions here will be drawn out *heuristically* by discovering how time *must* be to accord with what we understand and experience, rather than analysing the aspects which can be quantified and measured. To help avoid slipping into conceptual analysis, which must necessarily generalise and remove the individual context from experience, our arguments will often be analogical, rather than logical "The means whereby to identify dead forms is Mathematical Law. The means whereby to understand living forms is analogy" (Spengler:1932 Part 1, 4) We can never really enter inside another person's head, transport ourselves to another time, imagine how life would appear to a dolphin, dog or cat, but we can gain by insight, drawing analogies from introspection of our own selves.

Against the habitual use of the written word to objectivise, the aim is to express, albeit uneasily, a non-conceptual or pre-conceptual time, the time of the immediate present, a time as lived individually, not a time reducible to its spatial analogue. In philosophical terms this is close to Heidegger's time as "being" something to do with the very make up of human life itself before it is something measurable. Heidegger is not the easiest person to understand in the first place, even less so when he is trying to express conceptually what is essentially pre-conceptual but nevertheless the following quotes illustrate how his philosophy corresponds with the approach adopted here.

"Only if both Dasein's everyday historicising and the reckoning with 'time' which it concerns itself in this historicising, are included in our Interpretation of Dasein's temporality, will our orientation be embracing enough to enable us to make a problem of the ontological meaning of everydayness as such. But because at bottom what we mean by the term "everydayness" is nothing less than temporality, while temporality is made possible by Dasein's *Being*....." (Heidegger: 1962, 423).

"What seems 'simpler' than to characterise the 'connectedness of life' between birth and death? It *consists of* a sequence of Experiences 'in time'. But if one makes a more penetrating study of this way of characterising the 'connectedness' in question, and especially of the ontological assumptions behind it, there is a remarkable upshot . In this sequence of Experiences, what is really 'actual' is that experience which is 'present-at-hand' in the current 'now ', while those experiences which have passed away or are only coming along, either are no longer or are not yet 'actual. Dasein traverses the span of time granted to it between the two boundaries and it does so in such a way that, in each case, it is 'actual' only in the 'now' , and hops, as it were, through the sequence of 'nows' of its own time. Thus it is said that Dasein is temporal. In spite of the constant changing of these

Experiences the Self maintains itself throughout with a certain selfsameness". (Heidegger: 1962, 425).

We are in a living world in which our feeling of time is relevant to our experience with its memories, hopes and fears, and its transience. This time is organic rather than mechanical, internal not objective, psychological rather than physical.

It is possible rather simplistically to divide our everyday awareness of time into time as *immediately experienced* before we reflect on it and time as *understood.* In practice it is difficult to maintain this distinction because it is doubtful whether there can be pure experience which is independent of any understanding. Also, in the case of time, any understanding beyond the present must necessarily be partly of a time as understood. Nevertheless, the distinction will be followed, though not rigidly. This is because it clarifies the fact that, broadly speaking, time as experienced is relative to our condition as members of the human species, whereas time as understood is, in addition, relative to our condition as part of a culture with its own understanding of the world, which we absorb from the cradle.

Time as Experienced by our Species

We are just one type of consciousness aware of a world, one particular microcosm with its own particular way of experiencing a macrocosm external to itself. Although inevitably we privilege our human form of awareness over other forms, as it is the only one we are conscious of, there is no justification for believing that our view is any more ' true, than any other view. In other words what we experience as time is relative to us as a species, with our specific physical and psychological characteristics. Our time sense differs from other species through our brain structure, relative size, lifespan and organs of sensation, though we can never be certain precisely *what* these differences are.

The challenge is how to step outside of our own experience to identify what is uniquely human and to appreciate how time might be experienced by other sentient beings. Here two methods are used to tackle this obstacle. Firstly we can become cognisant of phenomena which are beyond our experience, but which objective or subjective *analysis* or use of material equipment which increases our powers of perception convince us must exist, even though we cannot experience them directly. Thomas Nagel makes a valiant attempt to do this in his essay *What is it like to be a Bat* (1974). An alternative approach is to proceed by *analogy*, though a synthesis of these two methods appears to have much to recommend it.

Using objective analysis first we can arrive fairly easily at a summary of the conditions in which the human feeling of time takes place:

1. First and foremost awareness of time requires consciousness:
2. This consciousness exists within a being of a particular physiognomy and brain.
3. It is linked to a perceptual apparatus which can only pick up stimuli within a certain circumscribed range. For example we see only a range of colours or hear a range of sounds. From the point of view of experience of time of the most relevant stimulus range is that of motion, as this is how we measure time in the external world.

Pure objective analysis cannot go much further but Ouspensky uses a more subjective analysis in his comments on our awareness of motion, motion being the most obvious spatial phenomenon which requires a time sense. There are for him four qualitatively different kinds of motion:

> "1 Slow motion , invisible as motion, for instance the movement of the hour-hand of the clock.
>
> 2 Visible motion
>
> 3 Quick motion, when a point becomes a line, for instance the movement of a smouldering match waved quickly in the dark.
>
> 4 Motion so quick it does not leave any visual impression, but produces definite physical effects, for instance the motion of a flying bullet" (Ouspensky: 1934, 434)

It may be argued that these distinctions are all too subjective to be of any benefit. But when the subjective is universal, although undemonstrable objectively, its truth can still be accepted and even be considered 'scientific'. Antonio Damasio holds this is true for consciousness as a whole (which is intimately bound to awareness of time),

> "The fact that mental images are accessible only to the owner organism....may cause some worry to purists raised on the idea that what another person cannot see is not to be trusted scientifically. Whether one likes it or not, *all* the contents in our minds are subjective and the power of science comes from its ability to verify objectively the consistency of many individual subjectivities." (Damasio: 2000, 83).

Ouspensky's different types of motion correspond to different subjective awarenesses of time as follows:

1. Slow motion is our awareness of the passage of time. Its passage is not directly felt but understood using our human memory and using our reason to compare the past with the present. This is 'memory' time. How far other species have a memory will determine whether they are aware of this.

2. Visible motion corresponds to what is immediately present. This is 'perceived' time.

3. Quick motion is when the visibly present is an object moving too fast for our eyes to physically separate it from the previous immediate present "The eye has a strange capacity of "remembering" for about 1/10th of a second what it has seen; if a point moves sufficiently fast for the memory of each 1/10th of a second to merge with another *memory*, the result will be a line" (Ouspensky: 1934, p436). This is a blurred form of perceived time where the immediate present cannot be isolated from the past and future.

4. Lastly is the type of motion we are aware of it only by its effect. It is a time never felt but understood by our reason. This is really a type of 'conceptual' time, using causal analysis to interpret how an event occurred.

Neither subjective nor objective analysis however can be of much help in understanding *animals'* time sense. Here both Thomas Nagel and Ouspensky resort to *analogy*.

Meanwhile, we shall return to the human species and the relation of motion to time. There is one other speed, which is seen as the limiting factor for human observation of motion – that of light. According to relativistic physics the speed of light is a constant that cannot be exceeded. This will be discussed later in more detail. Now however it is enough to state that this is intuitively persuasive. Light is the medium in which any visual perception must occur, so our particular means of seeing is unlikely to be able to work beyond the speed of the medium by which objects are presented to sight.

However, our time-sense exceeds its perceptual spatial manifestation as motion, or any ex-post facto conception of this motion as 'change' (whether this be change of position, intensity, size or colour) which interprets these perceptions of motion under concepts. Awareness of motion in space alone is insufficient to describe our sense of time. A whole range of our time sense is without any immediate visual perceptual stimulus. This applies above all to most our awareness of future and past time. Time awareness depends not only on our perception but also on our memory and reasoning as shown in the discussion of Ouspensky's four types of motion above.

Getting beyond awareness of time based on visual perception is difficult. To do this we have to change from using analysis (objective or subjective) to using analogy. Analysis from cannot be used when we move away from the perceptible to the imperceptible. To start with Ouspensky's analogy of a slit is examined. Unlike space where all is present simultaneously to our perception we are limited to a certain 'slit' of time which we call the 'now', what we call 'at the moment'. This slit is what we seem to share with other animals, though the 'size' of it may vary. The estimate of the human 'now' has been looked at by many psychologists and physiologists. William James devotes an entire chapter to this subject, the number of seconds of the 'specious present' being generally estimated as a maximum of 12. (James 1950: Chapter XV).

Of course, humans can conceive beyond this 'slit' using their reason and memory. But imagine the *perceived* slit is increased for certain perceiving beings or perhaps certain human beings with a 'heightened' consciousness:

"This perception will be able to grasp as something simultaneous, i.e. *as one moment,* all that *for us* takes place in a certain period of time, a minute, an hour, a day, a month. Within the limits of *its moment* such a perception will be unable to separate *before, now* or *after,* for it, all this will be *now. Now* will expand. "(Ouspensky: 1981, 34).

Such an expansion would affect not only our present visible time but also our kind of memory and our sense of causal relationships. In extreme cases, our whole life could be just a moment. In the face of the infinite magnitude of time and space this feeling of the littleness of our existence is not uncommon and with it often comes a quasi-religious numinous feeling.

> From our birthday, until we die,
> Is but the winking of an eye;
> And we, our singing and our love,
> What measurer Time has lit above,
> And all benighted things that go,
> About my table to and fro,
> And passing on to where may be,
> In truth's consuming ecstasy,
> No place for love and dream at all;
> For God goes by with white footfall.
> I cast my heart into my rhymes,
>
> That you, in the dim coming times,
> May know how my heart went with them
> After the red-rose-bordered hem.
>   *To Ireland In The Coming Times* W.B. Yeats

From the point of view of the history of life, or in Yeats's case, human ancestry, our conscious existence is indeed a tiny moment. In that one moment the whole of our life would be contained as a minuscule part of a larger whole, but somehow we as that speck can become aware of what we are a part of. This has parallels with Jung's concept of 'a collective unconscious'. The individual ego is minimised, leading to a world-view primarily influenced by archetypes, where individual consciousnesses are not very distinguishable from an archetypal universal human consciousness.

Another way to approach an understanding of internal human time, closely related to the image of the slit, is by drawing analogies using dimensional parallels. Since the nineteenth century, writers, the most well-known being Charles H. Hinton, have imagined a world seen by creatures who can only perceive in two dimensions, sometimes called Flatland. Whether a two-dimensional existence is feasible is not that important. These writers draw conclusions about how putative two-dimensional beings would interpret the presence of one more dimension i.e. the third dimension.

Ouspensky gives a vivid example: "Let us imagine a wheel with multi-coloured spokes rotating through a plane on which lives a two-dimensional being. The movement of the spokes will appear to a two-dimensional being as changes in the colour of the line lying on the surface. The plane being will call these changes phenomena and, observing these phenomena, he will notice a certain sequence in them. He will know that the black line is followed by a white one, the white by a blue, the blue by a pink...The changing colour of the lines will, in the opinion of the two-dimensional being, depend on some causes to be found there, on his plane. Any conjecture as to the possible existence of causes lying *outside* his plane he will dismiss as utterly fantastic and absolutely unscientific. And this will be so because he himself will never be able to visualise the *wheel* i.e. the different parts of the wheel on each side of the plane. Having studied the colours of the lines and learnt their order, the plane being, on seeing one of them – say, the blue one – will think the black and the white have already passed, i.e. have vanished, have ceased to exist, have *receded into the past*; whereas the lines that have not yet appeared - the yellow, the green and so on, and among them the *new* white and the *new* black which are to come – do not yet exist but lie in the future....This is how the two dimensional being arrives at the *idea of time* We see this idea arises from the fact that, out of three dimensions of space, the two-dimensional being is aware of only two; the third dimension he senses only through its effects on the plane; therefore he regards it as something distinct from the first two dimensions of space and calls it *time*." (Ouspensky: 1981, 49)

By analogy, for three dimensional species such as ourselves, it is the next dimension up, the fourth dimension which is our time. We only sense it through its effect on our three dimensional world. The two dimensional being mistakenly regards motion in the third dimension as time.

Another characteristic of dimensions is that we cannot see the whole of the dimension we are in. In our case, being in the third dimension we cannot see all of a three-dimensional object, say a cube at the same time. To see a cube in its entirety simultaneously we would somehow have to transcend our dimension and be in two places at the same time or in two times at the same place. This is mind-boggling stuff. By analogy again a line is a point extended in one dimension, a surface is a line extended into two dimensions, a cube is a surface extended in three dimensions and the next step is that the cube might be extended into four dimensions. If this fourth dimension is our time the extension of a cube can be symbolised but not actually seen in three dimensions. Just as we can represent the third dimension in two dimensional paintings using perspective so we can represent a three dimensional object extended over time. These representations for cubes are called hypercubes or tesseracts and trace the 'trail' of a cube moving through time. Hinton tried to extend his perceptual abilities by imagining cubes like this.

As a general principle, for a sentient being the next dimension up – the one of which the whole cannot be perceived, but only a 'slit' of it, is experienced as time. We are three dimensional beings but it is possible (just) to conceive of four-dimensional beings for whom our time is experienced as their space within a fifth dimension. They would 'see' everything that occurs in our time simultaneously. Presumably they would also regard beings such as ourselves who operate on less dimensions as having a limited understanding.

But rather than just use analogies which can't be 'cashed', is there a way of getting beyond our 'dimension'? In later chapters ways in which this may be possible are examined. But, for the time being we shall continue our review of time as experienced and understood by normal human consciousness!

Having said this, the state of normal consciousness is by no means uniform. In addition to individual variations, how time is experienced is also affected fundamentally by other physical factors such as our age, gender and genetic differences in our brain structure. Environmental factors are examined in the following section

Starting with age, as William James states, a well-known phenomenon is that "the same space of time seems shorter as we grow older" (James, 1890, 625). For a child the experience of the events of the day are so filled that a month before seems like a long time ago, whereas in old age that same month could seem like a few days in total. Even outside old age and childhood the subjective feeling of the passage of time varies according to whether we feel life to be full or empty. Waiting for a late train on a cold Winter's night or listening to a boring speech each minute can seem interminable, whereas in a snatched meeting with a lover every second passes too quickly. Our feeling of the passage of time is not measured by the clock but is measured internally without reference to the speed of the external world.

Women with menstrual periods, pregnancy and childbirth are more in touch with a cyclic natural time harmonising with the moon rather than the highly masculine clock time which treats natural cycles as intrinsically irrelevant. "Time has always been a highly genderised concept; linear time is phallic, male in shape, cyclical time is yonic and female in shape" Griffiths: 1999, 132). Any complete explanation of time must be able to cope with cyclical time. The issue of women who live closely together synchronising their periods is as much entitled to be included as theories of relativity. The dominance of 'masculine' time has become particularly marked since the 'Enlightenment'. It might work well as a useful working hypothesis for theories of classical mechanics, but is denied immediately in any life form by circadian rhythms. It cannot be extended as a definition which covers all temporal experience and in the present century is being rejected, even for some of the mechanisms of the physical world.

There are many other sensations of time which vary with our physical or mental characteristics independently of any visual perception. Examples are time in dreams, time in imaginative and creative activity, time as experienced in states of heightened consciousness (as well as states of damaged consciousness), sensual time – among many others. These, unlike visual perception, tend to vary more according to our environment and culture rather than the physical characteristics common to all mankind.

Time as apprehended in our environment

It is evident that our understanding of time is profoundly influenced by our environment. The awareness of time of a New York stock market dealer and an Amazonian Indian must be profoundly dissimilar. The 'environment' within which we live is culturally specific, in both a broad and narrow definition of the word culture. The broad definition includes the general environment in which we are reared and live in. Like all organisms this kind of culture affects our growth and reactions to the world; in human terms this is generally covered by the discipline of anthropology and could be named *'anthropological'* culture. Within this there is a narrower definition of culture. This is the world-view and tradition of the people to which we 'belong', which will determine how we conceive of the world, our value-system, beliefs, and practical training and skills. To give this a form we are given an interpretation of the past presented as a history and often our individual and people's future. As such, memories and expectations, fears and hopes are culturally influenced. Thus a culturally specific understanding of time is arrived at which could be named *'societal'* culture.

Oswald Spengler distinguished these two types of culture with the value-loaded terms 'primitive' and 'higher', but the distinction is in essence the same. In 'anthropological', Spengler's 'primitive' culture, any feeling of historical development is absent. Such peoples are ahistorical in the sense that although they may have long traditions and a high level of culture this is changeless. What happened 100 or 1000 years ago may be preserved in oral traditions but essentially there is no sense of growth or change. Disasters or successes such as a devastating drought or victory over an invading force are captured in myth.

Experience

There is no intention here of privileging a 'societal' time sense over an 'anthropological' time sense – rather the contrary. Not skewed by any societal preconceptions, in some ways anthropological time may be a purer time than those of 'advanced' societies. A truth for us as part of the human species, as expressed by this time sense, is much closer to the kind of time-sense which reveals itself later in this book. In this past and future are not cut off from the present but are porous and overlap. A powerful, but by no means unique example is Australian aborigine dream time:

"Dreamtime merges past, present and future... the Dreamtime or *Alcheringa* is a sacred time, a Great Time, qualitatively different from ordinary time and while the Dreamtime sustains the present, the present, in turn sustains the dream time myth" (Griffiths: 1999, 480)

What definitively distinguishes anthropological culture and societal culture is the creation of a written script. This is not necessarily the 'cause' of the emergence of societal culture but may be a response to a felt need to build on past knowledge which is no longer static and has a life of its own and a kind of destiny, and therefore a sense of history. As Spengler puts it: "'Historical' man, as I understand the word and as all great historians have meant it to be taken, is the man of a Culture that is in full march towards self-fulfilment. Before, this after this, outside this, man is *historyless*; and the destiny of the people to which he belongs matter as little as the Earth's destiny matters when the plane of attention is the astronomical and not the geological" (Spengler 1932 vol 2 p48).

Without going along with Spengler's bias or his list of cultures, the metaphor of their Springtime, Summer, Autumn and eventual death, there is nevertheless a useful idea from the point of view of our experience of time. This is Spengler's question: 'for whom is there history?' The answer must be for those who are full members of a societal culture. As full members they would have to be literate and educated so that their understanding of the world would include their understanding of time. The cultural understanding of time permeates deep into all layers of a culture, affecting how we view life and death, how we understand the past and future of the world and how we feel about what each day should bring.

Indian culture is still, despite Western influence, imbued with the Hindu notion of immense cycles of time which eventually repeat themselves. Islamic and Western cultures have a more linear attitude using calendars based on the birth of Mohammed and Jesus Christ respectively. The Chinese, on the other hand, have a strong but totally different cultural take on the passage of time, reverencing the past but not forming the kind of sense of evolution and progress found in Western and Islamic world views:.

"...In China the past had a definite social purpose, its use depending essentially on the concept of Heaven to ensure continuity in a world of political change....despite acquiring a mass of archival material stretching over a long period of time, the Chinese never developed anything corresponding to the modern Western concept of history....the European past, with its record of interaction between conflicting civilisations, religions and cultures lacked the unity and 'all embracing certainty of the Chinese and so presented historical problems of a kind the latter never encountered." (Whitrow: 1988, 90).

At this stage, before looking at time in the currently predominant culture which is the Western, different layers have been distinguished in the makeup of our sense of time. At the bedrock shared by all without differentiation is what stems from the biological make-up of our species. Sitting on this are the influences of the specific environment in which we were raised. On top of that is a third layer – the societal effects. All these layers can be coloured by one's personal history where our experience of time becomes part of our character. My own past is locked up. Before my mother left when I was seven years old I have suppressed nearly all memories. That, perhaps, explains, in part, why I am writing about time.

Time and Western Civilisation

The way the conception of time has evolved in the West could be the subject of a long volume in its own right. In the present context the aim is simply to summarise it, without a complete comparative historical account about how the Western view of time is distinct from others.

The first point to make is that the Western concept of time is its own creation and differs from others in many fundamental ways. Two of the most prominent are the assumption and use of mathematically regular clock time and the West's belief in the possibility of finding laws of nature by which what happens in the past determines what will happen in the future, in two words: *regularity* and *causality*. Other cultures' time may be less regular and closer to the natural rhythms of nature as witnessed by movement of the stars and seasons or within our own bodies such as sleep cycles, menstruation or ageing. As for causality, outside of 'educated' classes, despite the apparent hegemony of Western concepts there is still a strong attachment to the influence of chance, fate, destiny and luck.

The West is a crucial influence in two respects. Firstly, its cultural bias towards everything, including its understanding of time, is so dominant it is increasingly confused with what is 'objectively' accurate, rather than the views of a particular culture at the peak of its material influence. Secondly, the unequalled world power of the United States in recent years has added muscle to this intellectual dominance.

To understand how time is understood by Western man today it is necessary to reinstate a scepticism about fundamental assumptions of 'modern' Western thought established at the beginning of the 'enlightenment', which are now generally accepted without question as universal truths.

Bacon gave us the experimental method in which observation, repetition by objective experiment and the use of induction is the way to establish the truth. Following Bacon, time is henceforth externalised in that everything must be experimentally repeatable, the same cause leading to the same effect.

Descartes gave us two other basic requirements for scientific rules of analysis:

1. A problem must be broken down into smaller ones.

2. An argument should proceed from the simple to the complex.

Following the principles of Descartes and Bacon, for time to be used in scientific analysis it must be treated as homogenous, divisible and measurable. Time therefore becomes a variable that can safely be ignored because it is a constant which does not affect our knowledge of the world.

For understanding the natural world this was a healthy reaction to a great deal of earlier natural science which did not depend on induction from observation and on experiment, but relied on deduction from logical principles, regardless of verifiable evidence. Post-enlightenment scientific methodological principles are now regarded as essential when drawing conclusions about material reality. Theories, crackpot ideas, hypotheses reeking of dogmatism, prejudice or just plain ignorance cannot be granted the respectability of scientific status. One of the first tests before any attempt to see whether the findings can be replicated experimentally is: do they pass Karl Popper's 'empirical falsifiability' test? If there is no way a theory which purports to be scientific can be falsified by empirical evidence it cannot be 'scientific' (1963: Popper, 197).

However, where mind cannot be isolated so neatly from matter the experimental method cannot be applied so straightforwardly. If something, like a great mental revelation (like Newton's watching the apple fall and discovering gravity), is unrepeatable because it is unique and irreversible, you cannot find out how it was accomplished by applying Bacon's experimental method. Similarly, to anything which is more than the sum of its parts or indeed has no parts at all, Descartes' analytic principles cannot be applied.

Nevertheless scientific principles were employed with success in functionally developing our understanding and exploitation of nature. The error has been to extend them to cover all areas of knowledge, even those whose truth could not be confirmed by physical experiment and where the influence of consciousness and internal time could not be disregarded. (To be fair to Descartes he believed his method should only be applied to the external and not the internal world, which latter should be studied by introspection.)

Underlying the experimental method is a fundamentally dualist attitude. As long as matter and mind can be firmly separated, the approach will work, particularly for the material world. But the balance is firmly tipped on the side of the material rather than the mental. The world of mind is downgraded, leading to the Western leaning to see mind as epiphenomenal, a mere by-product of neurophysical processes. Hume's denial of the reality of the self led to the ultimate in Western attempts to eliminate any active role for the mind. In 1749, in Observations on Man, when David Hartley tried to link all actions of the mind with neurophysiological activity, he was interpreting Hume's philosophy and Newtonian mechanics into a materialist psychology. This over-simplistic view echoes down to the twentieth century particularly in psychology. Skinner's behaviourism and cognitive psychology are two examples.

Western 'materialism' is often taken to refer to the lack of any spiritual influence in a society greedy for material goods, but its origins go back to a world-view dominated more and more by the physical. Medicine is a paradigm model of the weaknesses and strengths of the Western approach. The contrast between the use of the inductive method when applied to human physical and mental faculties presents a startling divergence.

One of the uncontested achievements of Western science has been in medical care. Instead of basing research and diagnosis on unobserved and untested theories of how human bodies function, the study of the human *body* in physiology has revealed very clearly how the body functions. This has led to a sophisticated aetiology applied to cure injuries and disease.

But with psychology, when the study of *mind* tries to follow the same method it does not enjoy the same success. The lack of any firm consensus on how to study the mind is well summarised by Fritjof Capra in a chapter entitled 'Newtonian psychology' in his book *The Turning Point* (Capra: 1983). The result is that there are as many theories of mind as there are schools of psychology, unlike physiology where the material nature and functioning of the body are universally agreed. Trying to unify such diverse approaches as cognitive psychology, psychoanalysis and exclusively drug-orientated treatments under a single scientific theory is futile. Clinical Psychology is not an objective science at all.

This is by no means to dismiss the insights of psychologists, only to dispute their claim to be objectively scientific. Although Jung's approach is deeply illuminating and philosophically astute, he believed psychology is unquestionably a natural science, with of course his own particular system being the correct scientific one: "Although it has been obvious for at least two hundred years that philosophy above all is dependent on psychological premises, everything possible was done to obscure the autonomy of the empirical sciences after it became clear that the discovery of the earth's rotation and the moons of Jupiter could no longer be suppressed. Of all the natural sciences, psychology has been the least able to win its independence." (Jung: 2001, para 346)

The results of the discrepancy between the treatment of a diseased *body* in medicine and the treatment of a diseased *mind* in psychiatry are striking. Western medicine cannot get to grips with consciousness. Indisputably real human attributes such as grief, love, hope, despair, retribution or imagination, to name some picked randomly, are not capable of measurement, being intensive qualities, and not data susceptible to quantitative analysis.

At any time an influence from the past can spring up, often as the emergence of a memory, or when a perceived future influences a real or imagined anticipated event. Take the case of memory. While neuroscience can pinpoint the neural channels used in memory it cannot even begin to explain how they surface into consciousness and how they were 'stored':

"We have a solid corpus of research on factors governing learning and retrieval of memories, as well as the neural systems required to support and retrieve memories. But direct, conscious knowledge we do not have. We experience the contents that go into our autobiographical records – we are conscious of those contents – but we know not how they get stored; how much of each; how robustly; how deeply or how lightly. Nor do we know how the contents become interrelated as memories and are classified and reorganised in the well of memory; how linkages among memories are established and maintained over time, in the dormant, implicit, and dispositional mode in which knowledge exists within us. And yet, while we do not experience any of this directly we do know a *little* about the circuits that hold those memories. (Damasio 2000: 226-7 my italics).

The dualism in Western science including Western medicine is becoming weaker, as it becomes increasingly difficult to isolate mind from matter. A gap is opening in the scientific community between those who believe dualism is being replaced by a monism of material forces and those who believe that mind, though inextricably linked to matter, is not reducible to it (and yet at the same time want to abandon a dualistic approach).

The social or the 'soft' sciences try to obtain the same degree of 'objectivity' as the hard sciences and miserably fail. They can be functionally useful in using statistics to predict behaviour which can be used to allocate resources appropriately. But, unlike the 'hard' sciences they provide no objective truths about the nature of their field of inquiry, which is the workings of human thoughts and emotions. With the social and life sciences, the insights of the enlightenment became, in the late nineteenth and early twentieth centuries, an unchallenged dogma applicable to everything.

This dogma seemed to reach its apogee with Positivism, which despite being generally repudiated, still has a large following in the Great Britain and the United States. The youngest and most powerful of the Western states seems to drawn to the most extreme manifestations of Western thought patterns.

Positivism applies the principles of classical Western physics and natural science to social studies, indeed to everything. All details of human experience can be expressed in quantifiable measurements or all else they are not valid details. This data is collected and 'objective' laws of human activities such as economics, teaching and learning, or psychological behaviour can be developed which are as valid as the physical laws that have been developed in the natural sciences.

Just when it seemed that positivism had been finally discredited a new attempt to reduce all human activity to quantifiable data is being promoted by Richard Dawkins, with his concept of 'memes'. These are 'units of cultural transmission', anything from snatches of songs, stiletto heels to the idea of God, which are the atomistic building blocks which restructure the human brain. A new scientific discipline of 'memetics' will describe our cultural nature just as genetics does for our biological nature. This new breed of positivism has been enthusiastically adopted by other die-hard reductionists such as Daniel Dennett whose work will be discussed later.

Mary Midgley, the moral philosopher, forcefully points out that doctrines like memetics are coming just at the time when we seemed to be transcending the 'dead hand of behaviourism'.

She sees memetics as an "excellent illustration of the mess that tends to result when models drawn from the physical sciences are drafted in without good reason to explain human behaviour". (Midgley: 2001, 76)

Despite a nascent scepticism, as orthodox religion has declined science is taking its place as the new dominant unquestionable faith. This is why an examination of Western man's view of time becomes essentially the scientific establishment's view of time. Like all beliefs it is fundamentally irrational. This may seem an outrageous statement, given the benefits that the applied sciences of, say, medicine and engineering have achieved in bettering bodily health and in conquering the natural world. But the irrational element is not to do with the external world but with the underlying assumptions about the nature of the future and past of human culture and hence the 'purpose' of our lives. The notion of 'progress' is as irrational as fate or destiny which it replaces and is discussed in more detail later in this chapter.

The 'initiated' new elite of scientific high priests declare their truths as if they were gospel. In the case of social sciences, theories are nearly always refutable as the data collection methods and the statistical analyses results are always open to doubts about whether any data can truly be objective. Can they really be independent of the aims, preconceptions and cultural bias of the researchers? In the case of the 'hard' sciences untested 'working hypotheses' or overgeneralizations are being announced as truths. These truths however are contingent, subject to constant change and revision. However, as 'laymen', we are in no way qualified to dispute them.

"Scientists who insist that they are telling us how the world incontrovertibly is are asking for our faith in their subjective certainty of their own objectivity." (Appleyard 1993:54) The whole edifice is now tottering as it undermines its own foundations. This manifests itself very clearly in the collapse of the scientific paradigm of time, which, as we shall see later, is having to accept that an external time model, which is used in classical physics and relativistic physics, cannot explain everything, even in the physical world.

One would expect Western civilisation, with a highly developed historical sense and a tendency to objectivise, would have developed a profound sense of chronological and historical relativity. Instead it has created its own myth. This is made up of two assumptions.

1. Belief in the ancient–medieval–modern paradigm of history

In this view the achievements of Graeco-roman culture were buried in the 'Dark Ages' and then taken up again by the West. All other civilisations are simply ignored (India, Persia, South American civilisations), placed in a distant early historical background (Egypt, China) or muddlingly incorporated into this paradigm by treating them as religions rather than cultures (Judaism, Islam).

There is an unwillingness to consider seriously any ideas outside of the jejune Classical-Dark Ages-Western paradigm. In the case of time there is the telling example of the Big Bang theory. Even if it did happen, which is still disputed, there is still no answer to the question what happened before the explosion that created space and time. Fritzjof Capra points out how notions of a long term cyclic movement which some physicists are proposing could learn a lot from Hindu cosmology "This idea of time and space, which involves a scale of time and space of vast proportions, has arisen not only in modern cosmology, but also in ancient Indian mythology. Experiencing the universe as an organic and rhythmically moving cosmos, the Hindus were able to develop evolutionary cosmologies which come very close to our modern scientific models" (Capra: 1976, 219-220)

Interestingly, from a post-scientific perspective attempts are being made to revive ancient Indian cosmology, in particular using the Vedic tradition to show *their* perspective on Western physics and cosmology:

"The thesis of this book is that the framework of modern physics is too limited to accommodate many phenomena that occur within this universe.....at the present time, certain assumptions of modern physics have been adopted by people in general as the very foundation of their world view. *These assumptions are incompatible with the Vedic world-view."* (Thompson: 1990, p34 my italics).

But, do they need to be incompatible? The Vedic time calculations are on the same vast scale as modern astronomers. The cycle is as follows. A Satya-yuga is 1,728,000 years, a treat-yuga: 1,296,000 years, a dvapara-yuga is 864,000 years and our present kali-yuga 432,000 years. One thousand of these cycles make one day of Lord Brahma which is about four billion years. At the end of each cycle, known as a kalpa, the world is plunged into darkness and a new kalpa begins.

Edward T. Hall draws attention to the isolated Quiché Indians in the Guatemalan highlands. They have managed to retain the Maya time view despite the Spaniards almost entire destruction of the Mayan civilisation in South and Central America. The Quiché have "two separate calendars that mesh". One is civil and one is religious. (Hall: 1984, 83). "These two calendars interlock like two rotating gears to produce the Calendar Round, which only repeats itself once every fifty-two years." (Hall: 1984, 82). Also, totally at odds with our modern calendar, there is no beginning and no end, the progression of time periods being seen as cyclic rather than linear. Like many ancient peoples their calendrical calculations were superior to their European conquerors.

Support for different culturally based chronological views can also supported from the understanding of time among the American Indians, tribal beliefs in New Guinea, ancient China or even earlier civilisations such as the Chaldeans and Egyptians. Without a more agreed definition amongst physicists about the ontological status of time and reality and in the absence of a "theory of everything" physicists are divided among themselves. It is no use talking about quantum mysteries; the whole objective of the enlightenment was to remove mysteries, not to create them. In this confused situation to dismiss all approaches outside of the Graeco-Western tradition is short-sighted. Nevertheless despite these disagreements, as we shall see later quantum theory, already regarded by the lay public and many scientists as mystical in its own right, may still provide one of the keys for an inclusive paradigm of time.

2. Viewing history as progress

Paradoxically, physicists and mathematicians who work with a time reduced to space are the most enthusiastic believers in this notion of historical progress.

In contact with other peoples this notion of progress has led to an arrogance in which non-Western people who have not accepted and exploited our technological knowledge are seen either as underdeveloped or backward. No consideration was taken of how other cultures might view *us*:

Understanding Time

"When President Truman put forward the idea of the 'underdeveloped' world in 1949 – a world that must be industrialized, technologized, consumerized and indebted to Western banks and Western experts – what he didn't know was how underdeveloped other peoples thought Americans were, how disastrously immature the West was thought to be, in subtlety of communication, in social living, in earth-lore; how undeveloped in sensitivity, in sensibility, in reciprocity, in gift-giving, in kindness and in pity" (Griffiths: 1999, 197).

From a Western perspective, the world is progressing principally through the knowledge of Western experimental science and Western technology. The subtext is that the West is 'ahead' and in charge of this. Thus cyclic views of history, stasis or degeneration are never considered as serious options. Décadence is not a scientific notion, but it is often used to describe the West by many outside its main orbit. The similarities between the state of decadent Imperial Roman society and the modern West with their *panem et circenses* and collapse of traditional beliefs and values are many. Football matches resemble gladiatorial contests, Roman high-rise slums are like modern tower blocks and most of the citizens live on benefits in huge depersonalised cities.

So while relativity is allowed to apply to the physical world it is not applied to the historical context in which the theory was developed. From an historical perspective the thoughts and conclusions of Minkowski, Einstein or Niels Bohr are, in relation to the history of the universe, hardly drops in the ocean. In relation to human history they also lose much of their importance. We have been there before. The Babylonians, the Hindus, the Chinese as well as the Greeks were there. Oswald

Experience

Spengler believed every great civilisation has its own intrinsic conceptual view of the world 'reality' which is born, matures and dies with it, and this includes its science and fundamentally its understanding of time. The reason we attach greater importance to our own contemporary theories such as quantum physics is evidently because we are creatures of our time and of a culture whose emphasis is on the physical exploitation of its knowledge. But from an historically relativistic point of view they are no more valid than those of Parmenides or the dance of Shiva in Hindu cosmology. Indeed, if we examine the theories of great 'ancient' thinkers and sages and mathematicians, they are found to be as sophisticated and complex as those of great Western twentieth century theorists. Other cultures have taken on board the material fruits of Western science and technology without compromising their lifestyle and values.

Western civilisation has imprinted globally its view of how the world is and what time is and Western physicists are the arbiters of this view. But this view has evolved as well and is changing radically, often moving uncomfortably away from its own foundations. This evolution is discussed in detail in the chapter 'The New Physics and Objective Knowledge'. There is less and less justification for privileging Western culture's feeling for time as more 'true' than any other.

Just as gazing into the starry heavens at night can give us a sense of our littleness in space, so looking back at history can give a sense of our littleness in time, when we think we are on the verge of discovering a 'theory of everything'. But more importantly we rediscover the feeling of wonder revealing truths about the world. This sense of wonder self-evidently is not the copyright of physicists, astronomers and mathematicians. They certainly have no more insight into the "why" question and generally do not claim to have. As for the "how" questions, in classical science, and more so in the New Physics, "how" tends to mean what is mathematically consistent rather than an explanation of a causal mechanism. If something works mathematically and can be used functionally it is 'true'. This applies regardless of how many conceptual inconsistencies arise. Two obvious examples are of light being both a particle and a wave and more fundamentally the total irreconcilability of the two bedrock theories of modern science – quantum theory and relativity.

Where Western science is unassailable and brilliant is in dealing with the mechanics of how things work. But the price it has to pay is to resign hegemony, not only of the meaning of the world and consciousness to which most moderate practitioners would gladly accede, but also of their nature.

Individual Understandings of Time

How time is perceived and interpreted may depend on all the influences that have been discussed. But, hopefully, we are all unique and these influences will manifest themselves individually as they are refracted through our own personality.

We are alive in time and use our awareness of it every day, not just to practically navigate through our quotidian concerns, but also when we plan for the future, reflect on the past or simply enjoy the present! It comes naturally. We are born, grow, age and die in time. Time is inextricably linked to a sense of transience and our mortality. Different actions and states of mind give us a different sensation of time. Time spent looking after babies, sleeping, in traffic jams or waiting for buses, making love, in deep thought, or gardening feels different. Mood swings influence how we view time - when we are depressed time is a burden, when exalted it is overflowing and when we are content it flows equably. Different interests focus on different aspects of time. Views we have of time when we reflect on it will be a mix of all these.

Nevertheless, it is not difficult to disentangle our own views from a less individual general concept of time. We can go further than just isolating than our own personal bias. We can also get beyond the cultural preconceptions which form part of the more general concept. How much we can do this depends on our abilities to shed our preconceptions.

If we are passive receivers there is no urge to transcend the inbuilt structures developed from childhood; we remain content with a mindset determined by our physical and cultural conditioning. However, as part of an 'individuation' process we can emerge from these conditions just as Plato's freed captives emerged from the cave and saw how much their world-view had been just a reflection of their own environment. Part of developing our awareness of how we understand time and the world within time is to transcend as much as possible cultural and personal filters.

But is it really possible to arrive to at an unconditioned view, objectively valid and independent of cultural preconceptions? There is a temptation in this 'post-modern' age towards a view that all is relative to our inescapably culturally subjective modes of thought, so nothing has any 'absolute' meaning. But, cultural relativism does not necessarily mean there are no truths only that their expression will change. *Paradoxically, individuation leads to a less individual view, as beyond these filters lie glimpses of truths, of which those on the meaning and nature of time are fundamental*. These truths have the same form but directly that form is expressed its content will vary from culture to culture, as it must be actualised within the culture's expression modes. Applying this to time this is not limited to verbal concepts. How time is considered quantitatively in mathematics. How chronology is interpreted and how it gives individuals in the culture a sense of meaning and direction should also be understood. Post-modernism at least has a useful spin-off in correcting the arrogance of modern Western culture in considering its standpoint as universally valid.

The most important point is that our individual of sense of time can be developed and deepened. It is not an immoveable given. There would be no point in writing (or reading) this if it was otherwise. Our awareness of time, however, does not have to be so cerebral. There are other more direct ways of developing our sense of time apart from altering our mental understanding of it. In music and other art forms, the creative process itself and our response to it, makes us aware of a different sort of time. We can also develop a new feeling of time as an effect of emotional changes – feelings of love, ecstasy and detachment are examples of such emotions. Finally, how we use our body and physical action can change our feeling of time. These developments are intimately linked, as ultimately we are never just thinking, emotional, physical, or creative beings, but an indivisible combination of these. However, these are useful adjectives for highlighting different integral parts of the whole personality.

Chapter Three

Freedom

Conceptual thought alone cannot understand time as lived and experienced as the previous chapter has shown. It has a similar problem with action and we hit the old bogeyman of freewill. Akin to St. Augustine saying that he knew what time was until asked is Dr. Johnson saying: "We know our will is free, and there's an end on it." Both statements have a down to earth simplicity about them.

But, to repeat an earlier conclusion in a different context, thought is not the only way we are aware of the world. *Cogito ergo sum*, I think therefore I am, could only have been stated by a person who has unshakeable faith in the primacy of thought. Hence Descartes, who coined the phrase, saw animals as machines. But, we (and animals) are conscious, physical, sensual, sexual, emotional, intuitive and sometimes inspirational, as well as being the thinking beings we are when we put our cognitive jacket on.

Unlike the algorithmic external world where the same action leads to the same consequences regardless of when it takes place, in conscious decisions the timing can be crucial. What is right at one moment may be disastrously wrong at another. Thinking things out too reasonably can stifle action. Hamlet thought too much before acting and so was forced to act when he had little choice, doing the right thing at the wrong moment, with tragic consequences:

"Thus conscience doth make cowards of us all;
And thus the native hue of resolution
Is sicklied o'er with the pale cast of thought,

63

Understanding Time

And enterprises of great pith and moment
With this regard their currents turn awry,
And lose the name of action".
(Hamlet, Act 3 scene 1)

Any such intervention in the course of events by our decisions is ruled out by 'hard' determinism, consciousness being seen purely as an epiphenomenon, incidental to an event and in no way causal. In 'hard' determinism all events, including mental events, that are happening now are predetermined by past events, The future will also be determined by what is happening now and the traces of the past in the 'now'. Everything that will happen depends on what has already happened. So if we were all-knowing we could confidently predict the future and perhaps one day computers will be powerful enough to be able to do this.

From a scientific perspective, quantum theory in physics and chaos theory in mathematics have fatally weakened 'hard' determinism. Attempts have been made to replace 'hard' determinism by a 'soft 'determinism, where individual events are inherently unpredictable but their probability can be calculated. But this is ducking the issue, trying to avoid the unpalatable conclusion that the reason quantum events are indeterminable is because "quantum measurements interject our consciousness into the arena of the so-called objective world" (Goswami 1993: 98.) Where consciousness is not simply intervening in the objective world but intimately shaping it even such statistical probability is useless to anyone, except perhaps actuaries. We would use our conscious judgement or intuition to determine, for example, whether a close friend of ours is likely to, for example, appear as a witness in our defence at a trial despite threats to his life if he does so, rather than using any notion of statistical probability.

Freedom

If there were no consciousness there would be no freedom: "At any given moment all the future of the world is predestined... but it is predestined conditionally, i.e. there must be one or another future in accordance with direction of events, if no *new* factor comes in. And a new factor can only come in from the side of *consciousness* and the will resulting from it." (Ouspensky: 1981, 30-31).

Indeterminism is inseparably bound up with consciousness. But it does not follow that all conscious acts are free. Human actions need looking at a bit more closely and differentiating on a sliding scale. Some are definitely determined; others are less so. Some human beings, those not programmed into predictable responses, are more free than others.

So, not all action is free, but only the more important ones and not everyone uses this freedom. This is precisely the position of the French philosopher Henri Bergson in his book *Les Données Immediates de la conscience*, (translated into English interestingly enough, on Bergson's instructions, with the title *Time and Free Will*). Such a view of freedom makes no rigid distinction between free and determined acts: "freedom is not absolute, as a radically libertarian philosophy would have it, it admits of degrees." (Bergson: 1910, 166)

For Bergson, the determined self is the fragmented self where reactions, conscious or unconscious, relate only to the stimulus and not to each other. So long as we imagine a fixed intention such an explanation works; "the act follows the impression..; without the self-intervening." (Bergson: 1910, 168). But when the intention is questioned then the self becomes 'fluid' again, and the impression can be related more and more to the self. The deeper this goes, the more self-aware of our action we become.

Understanding Time

It has to be borne in mind that Bergson's self is not a fixed 'me', but something much more subtle, much more like the particular traits by which an artist's work is identified as genuine, than an analysable stable psyche attached to our brain. When free will is operating the antecedents are no longer fixed properties but will interact with the impression 'dynamically', the self changing during the action as well as the impression. The impression and the self integrate. In the limiting case:

"We are free when our acts spring from our whole personality, when they express it, when they have that indefinable relation to it which one sometimes finds between the artist and his work" (Bergson: 1910, 172)

This is the kind of relation which Bergson sees as distorted by a physical conception of self, where the present and future cannot blend with the past, altering and being altered by it. Bergson gives an example of this:

"It is by no means the case that all conscious states blend with one another as raindrops with water on a lake. The self...develops on a kind of surface, and on this surface independent growths may form and float. Thus a suggestion received in the hypnotic state is not incorporated in the mass of conscious states, but, endowed with a life of its own, it will usurp the whole personality when its time comes. A violent anger roused by some accidental circumstance, an hereditary vice emerging from obscure depths of the organism to the surface of consciousness will act almost like a hypnotic suggestion. Alongside these independent elements there may be found more complex series, the terms of which do permeate each other but which never succeed in blending perfectly with the whole mass of the self. Such is the system of feelings and ideas which are the result of an education not properly assimilated, an education which appeals to the memory rather than to the judgement...many live this kind of life and die without having known true freedom." (Bergson: 1910, 166)

Many philosophers, especially of the twentieth century Anglo-Saxon linguistic variety, like to address the 'problem' of free-will in terms of 'counterfactuals'. (Incidentally, there is also much current philosophy of science where counterfactuals are used to describe non-locality and theories of causation in quantum mechanics. This will be looked at later.) Hume's well-known "A person of obliging disposition gives a peevish answer; but he has the toothache" (Hume 1894, p88) can be expressed as a counterfactual "The person of obliging disposition would not have given a peevish answer if he hadn't had a toothache". This seems to makes sense. In contrast, as an instance of the level at which such *ex post facto* causal explanations of human actions in terms of counterfactuals cease to be plausible Bergson quotes an example from John Stuart Mill:

> "I could have abstained from murder if my aversion to the crime and my dread of its consequences had been stronger than the temptation which impelled me to commit it" (Bergson: 1910, 159)

Such a statement can be taken in two ways, either as a description of the agent's state of mind just prior to the murder, or as a realization after the event of forces within him of which he was not aware at the time.

If the murderer *had* been aware of these feelings at the time in the way he was afterwards, it would be improbable that his being conscious of these feelings would have made no difference to the relation between them.

The outcome was not therefore dependent upon these feelings seen as independent causes of his action, but upon whether and how much the murderer was conscious of them at the time. The explanation thus becomes 'I could have abstained from murder if I had known (more) what I was doing at the time'. To suppose otherwise would be to imagine that the only difference awareness would have made is that the agent would be given a spectator's view, a sort of ringside seat, as his aversion and dread were being overcome by the temptation.

By contrast, the pain of the toothache can be treated as distinct and independent from other feelings. Consequently, separation of the pain and the self experiencing the pain is simply accomplished, and it makes sense to talk of the pain as affecting the self and causing an out of character peevish answer.

A person whose actions can always be explained in counterfactuals is no longer capable of using his consciousness before he or she acts. In such habitual responses, as with conditioned reflexes, consciousness has gradually passed out of the action. Regularity of behaviour can often be a practical convenience. Obvious examples are typing or driving a car which leaves us free to pay attention to other things such as the text rather than the position of the keys, or the road conditions rather than how to change into fifth gear. As Stuart Hampshire writes "Precisely the point of having a firm fixed intention is that I do not need to think further about what I am to do "(Hampshire; 1999, 101). The question of whether such fixed intentions add or detract from our freedom of action depends on whether they serve us or we serve them.

To serve us, fixed intentions must remain related and integrated within the current state of consciousness, rather than becoming autonomous forces which elicit an unalterable response. "To be in a Passion you Good may Do/ But no Good if a Passion is in you" (William Blake, 'Auguries of Innocence').

Freedom

"Mechanisation, like *rigor mortis*, affects first the extremities...But it also has the tendency to spread upward. To be able to hit the right key of the typewriter 'by pure reflex' is extremely useful, and a rigid observance of the laws of grammar is an equally good thing; but a rigid style composed of clichés and prefabricated turns of phrase, though it enables civil servants to get through a greater volume of correspondence, is certainly a mixed blessing. And if mechanisation spreads to the apex of the hierarchy, the result is the rigid pedant – Bergson's 'homme automate'. First, learning has condensed into habit as steam condenses into drops; then the drops have frozen into icicles." (Koestler: 1976, 108)

Freedom, then, ceases when a person is inseparable from fixed habits and intentions. To develop oneself or to act morally are no longer possible. Stuart Hampshire sums this up as follows:

"Every rational man must know that his own interests are comparatively confined and that he habitually views situations, and his own performances, largely in the set and conventional terms that he has inherited and passively learnt. He knows that there are many ways of classifying his own actions which would never suggest themselves to him, and that there are many features of the world around him, which he never notices.

"He can come to understand the influences in his own environment and upbringing that have formed his habits of thought and his methods of classifying actions...He will count himself the more free in his thought whenever he is able experimentally to detach himself from his own habits and conventions of thought and to redescribe his own situation and conduct from a new point of view"...

"This policy of 'seeing round', and of testing an intention by alternative descriptions in any difficult case, is itself a maxim of conduct, a part of a particular moral outlook. It would not be adopted as a policy by anyone who claimed to be certain that only a limited set of descriptions are ever relevant in deciding whether something should or should not be done. This might be someone who was following explicit instructions which he believed to be God's instructions, or had for some other reason accepted a particular code of law as evidently binding on him. He might have decided that the whole of morality was contained in this code."

"For a man following a code of explicit and exhaustive instructions, moral issues would be matters of casuistry. He would be a type of fanatic, because only certain already listed features would be worthy of serious thought before action. (Hampshire: 1959, 213-216)

If the ability to detach ourselves from our 'present beliefs and intentions' is lost, then we must evidently be obliged to follow them. Hampshire's 'fanatic' becomes a passive victim of his habits and intentions in the same way as Bergson's 'homme automate':

"Our psychic states, separating them from each other, will get solidified; between our ideas, thus crystallised, and our external movements we shall witness permanent associations being formed; and little by little, as our consciousness thus imitates the process by which nervous matter procures reflex actions, automatism will cover over freedom" (Bergson: 1910, 237).

Such a state of affairs where there is no dignity or freedom left is ironically held up as an ideal by behaviourist psychologists, as exemplified by B.F Skinner who actually titled one of his works 'Beyond Freedom and Dignity'. Fritjof Capra neatly sums up behaviourism as "a psychology without consciousness that reduces all behaviour to mechanistic sequences of conditioned responses "(Capra 1983, p181.) Such a psychology is at a total loss to explain or treat any part of the human condition which touches on our inner selves and in fact denies that we have any inner self. On the other hand, it might completely capture a rat's psychology or help a person to cure arachnophobia,

By contrast, there are interesting parallels between Bergson's ideas and the conclusions of Nobel Prize winning neurophysiologist Sir John Eccles:

"But, further, if we assume that this 'influence' (of the discharge of one neuron) is exerted not only at one node of the active network, but also over the whole field of nodes in some sort of spatio-temporal patterning, then it will be evident that potentially the network is capable of integrating the whole aggregate of 'influence' to cause modification of its patterned activity that otherwise would be determined by the pattern of afferent output and its own inherent structural and functional properties...Thus, the neurophysiological hypothesis is that the will modifies the spatio-temporal activity of the neuronal network...The hypothesis is developed that these spatio-temporal fields of influence are exerted by the mind on the brain in willed activity. (Eccles: 1953, 276-285)

The resemblance of this scientific hypothesis to Bergson's philosophical theorizing carries over into the extent of free action: "It is not here contended that all action is 'willed'. There can be no doubt that a great part of the skilled activity devolving from the cerebral cortex is stereotyped and automatic, and may be likened to the control of breathing by the respiratory centres. But it is contended that it is possible voluntarily to assume control of such actions, even of the most trivial kind, just as we may within limits, exercise voluntary control over breathing." (Eccles: 1953, 271)

Understanding Time

For a conscious being anything can change now which no amount of knowledge or forethought could have predicted. In free acts our self works using an internal sense of time, what Bergson calls 'la durée', where prior to spatial externalisation past, present and future are not rigidly separated but can intermingle and effect each other. In extreme cases an act can change not just the present and the future, but the past (inasmuch as the only place the past lives on is in the memory).

In dreams, induced states such as hypnosis, dream-like states and narcotic effects on the mind, the moorings of objective reality are also broken. In this 'world' time is certainly not constrained in any linear order. There can be 'appointments in the past', often repeated in obsessive dreams. A huge variety of ideas and therapies tap on this source. Either by liberating us from some psychological blockage, opening and properly healing a long term wound or by bringing into consciousness a more complete 'rounded' self, our freedom of action can be enlarged. Psychological interpretations of time are a vast area of study not fully treated in these chapters. This is simply because a full discussion of the topic would open up a Pandora's Box, requiring a great deal of space, thereby swamping more limited aims.

We experience in time before we externalise these experiences We can only *think* about events *after* they have happened. Actions, moods, insights, sensations may have already occurred before we become conscious of the event.

Life is accomplished in time but conceived in space, where it loses its vital subjectivity. As Spengler puts it. "Time gives birth to Space, but Space gives death to time" (Spengler: 1932 vol 1, 173). Cognition is always *ex-post facto*, even though this may be a question of milliseconds. For neurophysiologist, Antonio

Damasio there is a lag between the beginning of neural events leading to consciousness and the moment one experiences the contents of consciousness as a mental 'event'.

The types of experience of time discussed here are not explicable in a 'spatial' time, where things happen one after the other. A reliance on spatialized time results in the belief that time is always linear and therefore quantitative and analysable into discrete moments. This is not the case.

So we come to a common sense conclusion that people are different. Some are freer than others or freer in different ways. Also the situation is not static: freedom can be increased or decreased.

In moral terms the more we can transcend the habitual reactions of a morality inculcated by rote, the more capable we are of becoming truly moral. In the last chapter we concluded by maintaining that once our cultural influences on understanding are transcended we are not left in a bleak post-modern world where there are no general truths. Similarly, once we can stand apart from our culture's moral code we are not in an amoral realm but rather in sight of a universal morality, which may be actualised differently in different cultures but is based on common tenets. However, its truths can only be acknowledged if our freedom is acknowledged and this freedom can only be possible if our internal sense of time is recognised, which can release us from the blind determinism of a purely objective world.

By realising our freedom the dignity which B.F Skinner wishes to dismiss is reaffirmed. We are no longer the tool of unquestioned prejudices or uncontrollable passions. "Autonomy is...the ground of the dignity of human nature" wrote Kant. That autonomy necessarily results in moral action as the self transcends its conditioning and moves towards acting from an internal motivation rather than reacting to events in a largely predetermined way. "A free will and a will subjected to the moral law are one and identical" (Kant: 1972, chapters 2 and 3.)

Eastern cultures, without the same dualistic tradition of mind and matter, express the same message but in a different way. The person rises higher to transcend without having to make a leap from the world of matter to the world of mind as they are not seen as being separate. In the Bhagavad Gita, what the West calls a free will is simply a will aligned to the Atman (Atman is the transcendent and therefore unchanging, true and infinite self that all persons and things possess. It is different from the apparent self of each creature which changes with time and space)

"Therefore, Arjuna, you must first control your senses, then kill this evil thing which obstructs discriminative knowledge and realisation of the Atman. The senses are said to be higher than the sense-objects. The mind is higher than the senses. The intelligent will is higher than the mind. What is higher than the intelligent will? The Atman itself. You must know him who is above the intelligent will. Get control of the mind through spiritual discrimination." (1954: Swami Prabhavananda and Christopher Isherwood, 49)

Chapter Four

Creation

"Everything may change in this demoralised world except the heart, man's love, and the striving to know the divine; painting, like all poetry, has its part in the divine; people feel this today just as much as they used to". Marc Chagall, quoted by A. Joffe in Jung's *Man and His Symbols*

Art can be the communicable representation of the divine using our freedom proclaimed in the previous chapter. It is an expression of our ability to transcend ourselves. This is true both for the creator and the audience. The feeling of pure unsubjected time is intimately part of this freedom for the artist when inspired and for listener or observer in their response. This is time of pure imagination lived and as expressed in art.

In whatever medium the artist works, in Langer's terminology a 'virtual world' is created. Sounds, materials, movements and words are employed to construct images of a true transcendental reality. For a sculptor a block of stone is transformed into an image, in literature and poetry words cease to be used to define but to evoke and in music and speech sounds are no longer meaningless aural sensations but meaningful form patterns. In each case, the material is subservient to the 'significant form' in which it is used.

But are these forms just ornamental illusions or do they convey a greater reality? We must recognise that works of art belong to a different domain. "Art is not the expression of a reality such as we see it, but…it is the experience of true reality and true life…indefinable but realisable in plastics. " (Langer, 1953: 81). Langer's virtual world is close to the philosophical 'noumenal' world:

Understanding Time

"For an artist the phenomenal world is merely material – just as colours are for the painter and sounds for the musician; it is only a means for the understanding, and the expression of his understanding, of the noumenal world...art goes further than ordinary human vision; consequently there are sides of life of which only art has a right to speak" (Ouspensky: 1981:133)

Literature on the surface is the art form most close to ordinary communication in that it relies on words, the medium for expressing concepts and logical arguments. But, here, just as in the non-verbal arts, a timescape is evoked in which objective time is overcome. In a well written work its scope and tempo are dictated by the writer, not by a page count.

Some writers are explicit about this virtual sense of time. Proust believed that art, and time as symbolised in art, were more real than anything else. He hesitated between writing a philosophical treatise and a novel to express his ideas. His dilemma was how to integrate a theory of art with the story of a life and in 'La Recherche du Temps Perdu' he triumphantly resolves this after having made various false starts in previous works.

When its narrator tastes the tea cake and his whole forgotten past comes flooding back as if it had been locked in the teacake, his past, present and future are dramatically changed. True, there is a kind of associationism but that would be a very shallow and incomplete way of describing the effects of the re-emergence of a lost past as an edifice around which his work emerged.

W.G.Sebald is another novelist in whose work the dilation and osmosis of time figure explicitly: "In fact, said Austerilitz, I have never owned a clock of any kind, a bedside alarm or a pocket watch, let alone a wristwatch. A clock has always struck me as something ridiculous, a thoroughly mendacious object, perhaps I have always resisted the power of time out of some internal compulsion which I myself have never understood, keeping myself apart from so-called current events, in the hope, as I now think, said Austerlitz that time will not pass away, has not passed away, that I can turn back and go behind it , and there I shall find everything as it once was, or more precisely I shall find that all moments of time have co-existed simultaneously..." (Sebald: 2001)

As the story grows more eerie, in a dimly-lit disused Waiting Room at Liverpool Station Austerlitz experiences something akin to Proust's bite of the Madeleine; one begins to understand, why even if our sense of passing time is unreal, it somehow protects us from the overwhelming flood of sensation of that would result if time could not be somehow spatially filtered out.

Poetry shares all the same characteristics of other works of art but is not covered in this review. It has, however, been treated separately at length in an as yet unpublished essay of mine: 'Poetry and Poetical Language: Love, Death, Dust and Time'.

Of all the art forms, music is the one most obviously connected with time. Music as pure time independent of space is virtually ignored by scientists and philosophers as embarrassingly outside the scope of their theories, which are supported by *visual* perception.

At the same time there are not many musicians or composers who are interested in philosophising or using scientific tools to explain their works. After all their expressive medium is sound not words. Victor Zuckerkandl, who admired Bergson, is an exception. He analyses and dismisses scientific studies which try to reduce all creation and appreciation of music to the influence of the sensation of physical phenomena of vibrations and stimulation of the ear and nervous system. In place of these, he argues that music can only be understood by extending to music the Gestalt principle in which "the individual part does not acquire its meaning from itself (or not exclusively from itself) but receives it from the whole".

In the chapter 'The Musical Concept of Time' in his book *Sound and Symbol* he describes four ways in which time in music cannot be physical externally measurable time:

1. *Time is not just a form.* It has content. Time is not just the medium through which events occur. The mere fact of the passage of time without any events occurring 'is not abstract empty form but a highly concrete experience. There would be no rhythm if time could not be experienced as such in itself' (Zuckerkandl: 1956, 203). This is what happens in the musical concept of rest. Zuckerkandl points out that although he uses high classical music as his example, it applies to all music and in the use of rest or musical pauses we find good examples in popular music. To take just a few examples there is a Doors song, 'Hello I love you', in which there is a pregnant pause before the music recommences. Similar effects are achieved in Sonny and Cher's "I got you babe' and in many compositions by Brian Wilson for the Beachboys. Often the effect is used as an envoi just before the end of the song.

2. *Time produces events.* Just as in the scientific theories of the infinitely small you cannot separate the observer from the body observed, in music you cannot separate the auditor from the sound heard: "The tones are the element that instigates, provides the possibility (of musical rhythm), they are as little in the picture as the listener whose

contribution to rhythm lies in the fact that he experiences it, not that he produces it. So we are left with only one series of data in which we can seek the source of the phenomena of rhythm: the *durations* of the tones. The answer to the question 'Of what are meter and rhythm the effects' can, then, only be that they are effects of the mere passing of time in the tones...only time can be the agent and source of the forces active in meter and rhythm" (Zuckerkandl: 1956, 206).

3.   *Time knows no equality of parts.* Here the distinction with physical time is much more easy to grasp. In clock time every portion (seconds, minutes or hours) necessarily has the same duration. "The equality of hours (in science) is the equality of distances travelled by clock hands "(Ibid, 209). In music, by contrast, keeping poor time is not "unequal length of the beats.. the commandment which is broken in a performance in poor time.. does not refer to the equality in lengths between intervals of time but to symmetry of mutually complementary wave phases" (Ibid, 211).

4.   *Time knows nothing of transience.* "The hourglass concept of time is incompatible with the simple facts with which music confronts us" (Ibid, 224) in a present "in which 'now', 'not yet' and 'no more' are given together in the most intimate interpenetration and with equal immediacy (Ibid, 227-8)

Zuckerkandl describes this time as organic time rather than subjective time to emphasise that this time is as real as physical time and not just a figment of our consciousness, i.e. purely psychological. "Every melody declares to us but that time anticipates itself "(Ibid, 235). It is immaterial and therefore not expressible within the scientific physical time, but it is nevertheless 'real'

Zuckerkandl's time is the internal time. This is apprehended prior to, or in the case of music, independent of any conceptions in space-time. Our initial response is to the whole, the whole being greater than the sum of its parts. It is an emotional aesthetic response by our whole being, not a cognitive act. If we start analysing we have lost it (exactly the contrary of science). It might be analysed in criticism or reviews afterwards about how the artists have used their technical skills to express themselves, but this is only after the whole meaning has been perceived non-conceptually. Gestalt psychologists maintained that a whole is much more than the sum of its parts; thus individual parts receive part of their meaning from the whole. Zuckerkandl when discussing the musical concept of time, extends the notion of gestalt to a temporal gestalt.

Zuckerkandl takes the position expressed in the introduction to this work that scientific time is correct, but it is just part of the story and without the whole story that part gets mistaken for the whole. "The Art can be the communicable representation of the divine using our freedom proclaimed in the previous chapter. It is an expression of our ability to transcend ourselves. This is true both for the creator and the audience. The feeling of pure unsubjected time is intimately part of this freedom for the artist when inspired and for listener or observer in their response. This is time of pure imagination lived and as expressed in art.

(Ibid, 247). One weakness, however, is that his views are drawn almost exclusively from Western music. There has been a recent resurgence of interest in the relationship of time and music, notably David Toop's work, which is much less centred around Western classical music and in tune with the growing enthusiasm for world music. Another shortcoming is that his views confined almost exclusively to music. But, any art that contains a time-element, and nearly all of them do, though not to the extent of music, can elicit Zuckerkandl's gestalt time.

The essential act of creation and appreciation of many art forms is an almost unconscious synthesis in time and space. According to Langer, artworks generally create virtual worlds of both space and time. For her, in the plastic arts space is the primary illusion; time is the 'secondary illusion'. Virtual time extends even into the most spatially dominant form of art, architecture, often referred to as 'frozen music'. When we look at a painting we are not examining strokes on a canvas but their part of a creation appreciated as a virtual whole.

It needs to be appreciated why artists feel their creations are part of or symbols of a more real world. It is a world of absolutes before division between the knower and the known, the creator and the created, form and content, in a word – before dualism. Merleau-Ponty in a very late essay 'Eye and Mind' wrote mainly about painting, but what he expressed can apply to any creative act. Between the physical actions of perceiving, painting, touching etc. and the world of nature there is no absolute boundary, both being part of the same whole.

"Immersed in the visible by his body itself visible, the seer does not appropriate what he sees; he merely approaches it by looking...the enigma which is my body simultaneously sees and is seen... it sees itself seeing; it touches itself touching; it is visible and sensitive to itself. It is not a self through transparence, like thought, which only thinks its object by assimilating it, by constituting it, but transforming it into thought ....it is a self through inherence of sensing in the sensed – a self, therefore that is caught up in things, that has a front and a back, a past and a future." (Merleau-Ponty: 1964, 159-190).

Understanding Time

Merleau-Ponty's insistence on the primacy of the physical in perception, tactile and aural, as well as visual, is wholeheartedly supported by neurologist Antonio Damasio. He argues that "There is no such thing as *pure* perception of an object within a sensory channel, for instance, vision...To perceive an object, visually or otherwise, the organism requires both specialised sensory signals *and* signals from the adjustment of the body, which are necessary for perception to occur. " (Damasio: 2000, 47).

When you perceive for instance, the approach of a car coming fast towards you as you cross the street, "The car approaches, and the position of your head and neck is altered as you orientate in its direction... Experience or knowledge of something, in a word, consciousness, comes later" (Damasio: 2000, 146).

This has similarities with Bergson's views expressed in the previous chapter but is much less cerebral in that the perception is with the whole body and all the senses, not limited to just a pre-conceptual consciousness, but to a pre-conceptual 'being in the world'. But this is an almost physical reaction which precedes Bergson's pre-conceptual consciousness, which he calls 'intuition'.

This state is in harmony with all because it is before self-consciousness which makes all relative to a microcosmic observer. At this stage there is no separation as there is no space or external time separating nature and events. In an essay much admired by, amongst others, Thomas Mann and Goethe, Heinrich von Kleist explored the art of puppetry as approaching this state of grace before knowledge. In their lack of affectation and self-consciousness they are like man before he eats of the tree of knowledge and is cast out of Eden.

Creation

"Kleist shares with Kafka.. the insight that it is only our concept of time which makes us think of the Fall of man as a historical event in the distant past. It is happening all the time. The biblical story is a mythical representation of constant human awareness of self and therefore of separation." (1994: Kleist, 10). This is an extreme view but describes with chilling accuracy what would happen if we remained in a preconscious state, where all is unity and there is no relativity. It is also, for Merleau-Ponty, close to the painter's view of the world who, when inspired, is so close to his work that "it becomes impossible to distinguish between what sees and what is seen, what paints and what is painted" (Merleau-Ponty: 1964, 167)

There are here a whole cluster of approaches to creativity and vision, which although they differ in emphasis and degree, share the following features:

1. Before conscious thought there is an immediate perception which can be felt and expressed.

2. In this awareness space and time are not rigidly ordered, yet are an inseparable constituent part of our experience, rather than external forces.

3. The self is not conscious or fully conscious and there is a unity which precedes division in self-consciousness between the knower and the known. Art "takes us out of ourselves".

What is true in this world are not the logical truths of identity or the experimental truths where an hypothesis is borne out by the evidence of our senses. What creative art or visionary insight is approaching is a different sort of truth in a different sort of time. The musician knows a note to be false; even though the timing may be objectively impeccable it sounds wrong. Keats' 'Beauty is Truth, Truth Beauty' is an attempt to express this.

The images created in art are based on our experience and so on our culture. Many of them appear to be untranslatable between different times and places. The world of an Athenian philosopher or playwright in Ancient Greece, a FDIC priest or Indian sculptor, an Enlightenment natural scientist or Romantic poet, seem so far apart that there can be little shared understanding. So are extreme postmodernist theorists right in denying the validity of anything universal, with attempts to do so being just a dangerous sign of cultural arrogance trying to construct order when none exists? "I have come to believe that the world is an enigma, that is a harmless enigma which is made more terrible by our own attempt to interpret it as though it had an underlying truth" (Eco, 1989 : 95).

To avoid this nihilistic conclusion there has to be a distinction between form and content. Content is always relative, subject to individual circumstances, but contained within a form, which is more durable. Goethe's famous line in *Faust* Part II that 'everything transitory is but a symbol' invites affirmation but poses the question: What is it a symbol of?

All cultures have had artists, poets and musicians who strive to express themselves with their own culture's symbols and shared outlook on the meaning of life. These have often lost their original meaning and power as the culture which created them died. Despite this they are expressions of a pure consciousness which rests eternally 'true' beyond the content of the expression which will necessarily be dependent on the nature or nurture of the artist.

In addition, these cultures also had sages, philosophers and often religious leaders who try to put these absolutes into words, often with a surprising degree of unanimity, as recorded in Aldous Huxley's *The Perennial Philosophy*.

What we are dealing with is not a postmodern 'enigma' or a metanarrative capable of being deconstructed into the historical and social circumstances which shaped it. These kinds of analysis may be helpful criticism or reviews afterwards about how the artists have used their technical skills and cultural background to express profound mysteries as truths. But this is only after the whole meaning has been created and perceived non-conceptually. In Art the artist seems to be expressing the workings of the imagination to create representations which resonate in ourselves as images of this truth. But this play of the creative imagination, although actualised by artists, is not limited to them. It is the source of all that is new in any field in which a deterministic, spatially extended world is not allowed to suppress the synthesis of feelings, thoughts or ideas that mingle in the creative act. Hence it doesn't sound strange to hear scientists, athletes or cooks, for example, talking about inspiration, beauty and creativity.

Nearly every field of knowledge can be approached in this fashion before analytical thought breaks down a unity by separating all out in space and time.

"Reason, system and comprehension kill as they cognise. That which is cognised becomes a rigid object, capable of measurement and subdivision. Intuitive vision, on the other hand, vivifies and incorporates the details in a living inwardly-felt unity.....The artist...sees the becoming of a thing....whereas the systematist...learns about the thing that has become" (Spengler: 1932 vol 1, 102).

Theatre has often managed to create a spatio-temporal unity in which the discreteness of space and time is suspended for the duration of the performance. It is clear that one sign of a good performance is that the audience enters not only into the visual and aural magic of the piece but into the pace of time within it. A sure sign of boredom is to look at one's watch: "Awareness within the audience of 'clock time' is strongest during any performance when it totally 'loses' its audience, causing the audience to check continually how long it is until the interval" (Trimingham: 2001, 201) Sound, space, movement, visual effects and dialogue combine to create a virtual time sense. Theatre can create the same kind of temporal gestalt as music:

"the unity of the work proves crucial in theatrical communication.....what has been largely unrecognised is how much this sense of unity depends on a shared sense of time with the audience, an infinitely flexible sense, a sense that ebbs and flows, expands and contracts, a sense that is structured by what the audience sees, hears and haptically senses; a sense moreover that is capable of holding within it the past, the present and the future: a truly temporal gestalt." (Trimingham: 2001, 205)

In modern performance a good deal of attention has been addressed to the relation of theatre to time. "The structure of history, the uninterrupted forward movement of clocks, the procession of days, seasons and years, and simple common sense tell us that time is irreversible and moves forward at a steady rate. Yet these features of traditional time were challenged as artists and intellectuals envisioned times that reversed themselves, moved at irregular rhythms, and even came to a dead stop. In the *fin de siècle,* time's arrow did not always fly straight and true" (Kern: 1983, 29).

Creation

Toward the end of the nineteenth century, parallel with the revolution in scientific concepts of time leading to post-Newtonian physics, new technology meant that artists could use newly invented media to subvert Newtonian time.

With the moving picture, whether television or cinema, an art form comes into being which, at first sight, would appear to be more capable than any other of representing time in all its manifestations. The illusion of continuous movement it creates mirrors the flow and direction of subjective time. In addition, with flashbacks, different speeds, sensitive musical accompaniments, cutting room techniques to highlight and suppress, it seems ideal for capturing the varying moods of consciously experienced time.

Bergson saw the cinema as providing a medium capable of representing human experience of time, anticipation and memory. It was left to French philosopher Gilles Deleuze nearly one hundred years after Bergson to apply his notions in detail to film. Instead of claiming to have a detailed knowledge of film theory, what follows leans very heavily on his work, selecting from it the directors with whom I am familiar.

Deleuze was an extraordinary man with close links with the French 'New Wave, participating in the alternative educational structures created in the May 1968 rising. He was a prolific writer. His subjects included traditional thinkers such as Hume and Kant, less mainstream philosophers such as Nietzsche and Foucault, and writers such as Heinrich von Kleist and Fyodor Dostoyevsky – all, incidentally, referred to in this book.

In *Cinema 2 the Time-Image* Deleuze explores how cinema can uncouple time from its spatial manifestations in movement, which were examined in his preceding volume on the 'movement image'. The time-image abandons a linear sequential narrative approach.

Understanding Time

"Time as progression derives from the movement image of successive shots. But time as unity or as totality depends on montage, which still relates it back to movement or to the succession of shots. This is why the movement-image is fundamentally linked to an indirect representation of time, and does not give us a direct presentation of it, that is, does not give us a time-image. The only direct presentation, then, appears in music. But in modern cinema, by contrast the time-image is no longer empirical or metaphysical: it is transcendental in the sense that Kant gives this word: time is out of joint and presents itself in the pure state. The time-image does not imply the absence of movement.... but implies the reversal of the subordination; it is no longer time which is subordinate to movement; it is movement which subordinates itself to time". (Deleuze: 1989, 271).

However, this use of the time-image was hardly noticeable before the 1950s and is still rare now, being kept out of the mainstream. In general cinema still relies heavily on the movement image being used to produce propaganda or mindless pap. Both objectively and subjectively film can be a sophisticated tool for distortion of the truth. Propaganda distorts the movement image as Leni Riefenstahl did for Hitler. Much less dramatically, I remember very clearly a pure distortion of which I was the victim. After working frantically on a project I initiated to restore a Regency open-air swimming pool in Bath (where Captain Webb used to practise for his Channel Swim) I was running out of time. The Summer season had begun and the pool was not yet ready for opening. With a huge last minute clearing up operation it was ready one Friday for opening the next day. We invited the press and television. I was interviewed against the backdrop of the Regency changing rooms bubbling with enthusiasm. Was I sure it was ready? Yes of course. In the meantime another camera crew was filming a huge pile or rubbish, which we got rid of later that night. When the interview was broadcast I was no longer in front of the glorious changing rooms but standing with the huge rubbish heap behind me.

As for the production of pap, Deleuze quotes Antonin Artaud 'The imbecile world of images caught as if by glue in millions of retinas' (Deleuze: 1989, 165). Instead of becoming a cinema being a series of images which "must produce a shock, a nerve wave which gives rise to thought" (Deleuze: 1989, 165) it risks becoming the opium of the people as they go on-line for their fix. With those provisos of the dangers of the moving image we shall look at Deleuze's time-image.

When time is delocalised from movement it means that cinema can capture what I have called internal time and Deleuze calls 'pure time'. Time is presented as it is in thought, reflection and emotion directly. When this succeeds Deleuze maintains that a 'virtual' image is created using the same word as Langer does to express how the arts use virtual creations as the "expression of a true reality and true life....indefinable but realisable in plastics" (Langer: 1953, 81). In the virtual time-image the recollected image merges with the present image to the point where they can no longer be distinguished. They are united in what Deleuze terms the 'crystal' image.

Every aspect of the film takes part in this decoupling, the sequence of shots, scenery, narrative, music, camera techniques, character presentation and story line. The most obvious example is the flashback. Deleuze however denigrates this as overused and not always achieving the desired effect. The book is crammed with examples of a great variety of techniques by a host of directors. The details below are limited to three,

Understanding Time

Orson Welles, Resnais and Visconti. Welles and Visconti need no introduction. Alain Resnais though is not so well-known. His most widely known film is 'Last Year in Marienbad' though this is not typical of his approach as it was a collaborative production with another French director, Robbe-Grillet

Orson Welles' best known film classic is 'Citizen Kane'. Deleuze describes this as "the first great film of a cinema of time" (Deleuze: 1989, 99). A series of subjective flashbacks occur in the film as witnesses recall the newspaper tycoon Kane after his death. All of these are, of course, recollections in the film's present. But by a collage of former presents Kane's life is portrayed as the questioner attempts to discover the significance of Kane's last word 'rosebud'. Within these flashbacks Welles further subverts any linear chronological time by use of cinematographic techniques. One example is his innovative depth of field; another is the use of mirrors. "Welles invents a depth of field...along a diagonal or plane, making elements from each interact with the rest, and in particular having the background in direct contact with the foreground.. as in the suicide scene where Kane bursts in through the door at the back, tiny, while Susan is dying in the shadow in mid-shot and a large mirror is seen in close-up.....the new depth... directly forms a region of time, a region of past which is defined by *optical* aspects, or elements borrowed from, interacting planes. It is a set of non-localisable connections, always from one plane to another, which constitutes the region of past or the continuum of duration" (Deleuze: 1989, 107-8).

Resnais goes a step further: "He obtains a generalised relativity, and takes to its conclusion what was only a direction in Welles" (Deleuze: 1989, 117). The present no longer intervenes as the centre and there is no one character or object, even one as tenuous as the 'rosebud' in 'Citizen Kane', to unify the fragments of the past in the present.

Resnais is a master of what Deleuze terms the 'noosign', an image which transcends itself and can only be interpreted by thought. Images are no longer connected chronologically in order of their succession but by thinking about the feelings they evoke. "Resnais goes beyond characters towards feelings and beyond feelings towards the thought of which they are the characters....if feelings are sheets of the past, thought, the brain, is the set of non-localisable relations between all these sheets (Deleuze: 1989, 125).

In his shots Resnais plays with the relations between images and different versions of the past. These could be one person over time as in 'Je t'aime, je t'aime', two people ('Muriel' and 'Hiroshima mon Amour'), more than two, or group memories, often interspersed with collective memories of different cultures.

Flashback is too crude a tool to capture this and Resnais uses it sparingly and, even when he does, it is unclear whether it is a flashback or not. Instead he links and relinks images, as in 'Je t'aime, je t'aime, where there is a return to the same image, but caught up in a new series. In 'L'Amour à mort' little feathers or corpuscles flutter at various speeds and in various arrangements in the night and keep reappearing throughout the film.

The primacy of the 'movement-image' where time follows the sequence of the shots is lost. Images are connected cerebrally as 'time-images': "the screen itself is the cerebral membrane where immediate and direct confrontations take place between past and future" (Deleuze : 1989, 125). Cinema with Resnais is capturing Bergsonian time where linkages between past and future are not mediated through space and motion.

Understanding Time

In a footnote Deleuze states that "Visconti is in a profound sense a film-maker of time" (Deleuze: 1989, 97). In contrast to Welles and Resnais the crystal of the merging of the recollected image and the present image is charged with the feeling of 'decomposition' of the recollected image. Visconti chooses his themes with this intention "everywhere the thirst for murder and suicide, or the need for forgetting and death". (Deleuze: 1989, 95). Whether it the dissolution of an old aristocracy in 'The Leopard', the slaughter of the S.A in 'The Damned' or the slow death of the musician in 'Death in Venice' any sense of hope is destroyed.... it is too late. The past is beyond redemption. "It is a vanished past, but one which survives in the artificial crystal" (Deleuze: 1989, 97). Against this Visconti sets the 'real' past, history, which is captured obliquely, never as background scenery. The feeling of too late of the 'crystal time' stands out from the static time of history: "the 'too-late' is not an accident that takes place in time but a dimension of time itself" (Deleuze: 1989, 96). The oppressive charged atmosphere which this juxtaposition creates requires Visconti's genius for portraying great scenes, use of colour combinations, pace and camera techniques to evoke it. Deleuze's 'too-late' time dimension is to close a poetical or theatrical tragic time. It is Lear with Cordelia, again 'too-late', the time which might have been which has become more potent than the actual present.

The images of these films "imply non-local relations. These are direct presentations of time. We no longer have an indirect image of time which derives from movement, but a direct time image from which movement derives". (Deleuze : 1989, 129).

Lastly, Borges should be mentioned as an outstanding example of an artist whose attitude towards time was central to his work and also an avid filmgoer and occasional film critic, particularly interested in films which used time. He believed that time was the central riddle of the world. With his eclectic knowledge of philosophical and esoteric works he arrives at riddles very similar to those posed by quantum theorists, which will be looked at in the next chapter of the new physics. Paraphrasing Schopenhauer he writes:

"No man has ever lived in the past and none will live in the future....we might compare time to a constantly revolving sphere; the half that was always sinking would be the past, that which is always rising would be the future; but the invisible point at the top, where the tangent touches, would be the extensionless present. As the tangent does not revolve with the sphere, neither does the present, the point of contact with the object, the form of which is time, with the subject, which has no form, because it does not belong to the knowable, but is the condition of all that is knowable" (1970: Borges, 268)

Borges wrote a review of 'Citizen Kane' describing it as a labyrinth without a centre. Deleuze's concept of cinema non-locality fits very easily with Bergson's view of time as flux 'la durée' and with the concept of non-locality discussed in the next chapter. By synthesizing these developments in philosophy, art and science a new integrative paradigm of time emerges, with 'non-locality' being one key word in a new vocabulary which can express this.

Chapter 5

The New Physics and Objective Knowledge

"If we identify Einstein's theory of relativity with the modern era of physics, then I contend that modern physics will not solve the riddle of time. But postmodern physics might. Two areas of investigation look promising. One is chaos theory, the other quantum mechanics." Paul Davies, *About Time: Einstein's Unfinished Revolution*. p.272

Although I have spent many years studying time I would never have written at length about it until pronouncements on the nature and significance of time in the media and the gush of books began to irritate me so much that I felt a riposte was necessary. They are always based on the views of physicists and astronomers, as if no one else is entitled to speak of their experience of time. It wasn't an irritation with the counter-intuitive conception of time they presented; there is something inherently baffling about time and any real understanding is quite likely to require new uncustomary ways of thinking. It was rather the assumption that the only real time is that used as a linear measurement of physical occurrences in the world or cosmos. If asked 'what is this time?' the standard conventional scientific response has been 'we don't really know, but for all practical purposes, it is its measurement'.

I could think of many aspects of temporal experience not conducive to explanation within this type of time. They range from an understanding of the development and growth of an individual or an historical trend, how original thoughts or works of art are created or interpreted, to natural phenomenon in living creatures such as the tendency of women living in close proximity to synchronise their menstrual cycles or the cyclic migrations of birds and fish.

Contemporary accounts of time ignore these, concentrating more or less exclusively on the impact of current theories in physics and cosmology. One of the purposes of this book is to place these in a wider perspective. The risk is that the sheer volume of scientific material on this subject if looked at in too much detail will distort the balance of this book. To avoid this only a selective general overview is attempted, focusing around the question of what aspects can be integrated with other non-scientific perspectives. Another reason for brevity is that this ground has been covered recently in books such as Mike Sandbothe's *The Temporalization of Time: Basic Tendencies in Modern Debate on Time in Philosophy and Science*.

The views espoused by physicists and cosmologists may be suitable for describing some of the lifeless mechanics of the external world, but have little bearing on phenomena where time needs to be treated as an active dynamic presence, rather than a passive backdrop. Indeed, they do not even apply to the whole of physics. As Nancy Cartwright (Cartwright: 1999), states physics is now a 'dappled world'. What works in one conceptual paradigm does not work in another. All that needs to be assumed is a *ceteris paribus* principle which, assumes that all the factors not taken into account in a particular paradigm being applied remain the same and do not affect the outcomes. Relating this dappled approach to the subject of time the identification of time with its linear measurement is correct in the realm of Newtonian science. But in relativity theory and, even more, in quantum mechanics and a possible quantum cosmology, it does not work. Instants of time can no longer be visualised as points on a line. In relativity they are part of a space-time continuum; in quantum theory distinguishing between an external measurable time and the conscious observer doing the measuring becomes problematic. Consequently the linear view is only valid in a limited part of a limited sphere.

While the achievements of the uses in science of a concept of time reduced to its qualities analogous with space must not be belittled, the danger lies when physical science regards this as a complete description of temporal reality. When it is viewed as revealing truths about the world, rather than being essentially a useful way of conceiving the world in order to use it functionally, it gives rise to a 'hard' realist interpretation, which Logical Positivists adopt. If you can't measure, and therefore analyse, a phenomenon it does not exist

Time is either not considered as a significant 'variable' or 'frozen' as its spatial analogue. Classical science can only handle time as spatially expressible. Therefore it has never been able satisfactorily to come to terms with aspects of time that cannot be reduced to spatial terms, namely its flow (which means you cannot dissect it for analysis) and its direction (which means you cannot reverse or repeat experiments or indeed events). Analytic truths use a logic which is only true if these essential attributes of time are ignored. The laws of identity and non-contradiction are only absolutely true in an unchanging timeless universe. 'A' may be not 'A' a nanosecond later.

The title of this chapter refers to objective knowledge to stress the difference between knowledge of the external world as objects, be they electrons, atoms, chairs or stars, and other ways of knowing the world. The extreme materialist position is that only the external world is real and that the mind is an epiphenomenon: so the only true knowledge is objective knowledge. Often 'natural' science rests on this point of view.

It is only the objective which is 'measurable' and can give quantitative results.

But this position can only be maintained if consciousness, above all consciousness of time, has no appreciable influence on the content of objective knowledge. The truths of science are universally applicable because time is not a variable, being a constant expressed by its spatial measurement. Time and classical Science are basically antithetical. "This principle of the irrelevance of the time may be extended to all scientific laws. In fact, we might interpret the 'uniformity of nature' as meaning just this, that no scientific law involves time as an argument (Russell: 1963, *The Notion of Cause, p149*).

Nevertheless, time conceived of externally as its measurement must always relate ultimately to a consciousness, which does the measuring. Measurement is not an integral property of time; it will pass without being measured. Internal time, 'mind-time; works outside any measurement. At the same time though, insights from it can only be actualised in the external world. In philosophical terms this is a variation of the Kantian dictum: "Thoughts without content are empty, intuitions without concepts are blind". The truths of pure thought, formulated in logic and quantified in mathematics are timeless and therefore vacuous – pure form without content. Pure reason has it limits, as Kant's *Critique of Pure Reason* established. One of his objectives was to limit reason to leave room for faith. But it is not only faith, but any apprehension which precedes or defies conceptualisation. This applies *a fortiori* to time.

The reality of pure consciousness, being without thought, is just content, a never-ending flow of impressions in time alone, an 'eternal now', with no external 'hooks' to convert them into concepts rather than impressions. But this conscious awareness of a being in time, blind and unformulated as it may be, precedes all possibility of understanding, whether this be conceptual or non-conceptual. Put in more philosophical terms, the manifold content of the raw data of sensations need to be filtered into spatio-temporal order, *with temporal order being the essential first step*, before they become intelligible.

In pure time there are no pauses from which to start a measurement. Heraclitus was aware of this in his famous statement that you never step in the same river twice. Conversely, Parmenides and his pupil Zeno of Elea argue for the impossibility of motion and change because they adopt an atomistic approach in which there is always a smallest unit at which one can pause. But time is inseparable from a conscious subject, because it does not belong to the object itself, but only its relation to a conscious perceiver. When attempts are made to treat it as an external capable of being broken down into atoms, paradoxes inevitably arise. Zeno's most famous paradox is the race of Achilles and the tortoise. The tortoise, who obviously moves more slowly is given a head start. But according to Zeno of Elea, Achilles, despite being 'fleet of foot', can never overtake the tortoise. In Zeno's atomistic time while Achilles runs the distance from where he is to where the tortoise started the tortoise will have already moved forward. Then when Achilles runs the distance to where the tortoise has advanced to the tortoise will have already moved forward again and so on ad infinitum. This paradox appears just to be denying the possibility of comparing relative speeds, but in fact it is denying the reality of motion by turning what is a continuous process into a discrete process.

Zeno's paradox of the arrow spells this out more specifically. At any instant of time an arrow in flight is at a certain point. For it to move time would have to pass but this would mean that the instant contains still smaller units of time which is not possible because an instant is the smallest unit of time and therefore indivisible. As the instant chosen is arbitrary this means the arrow must be at rest at any instant.

The truths of this type of logic can only operate in this unchanging dead universe of static moments and always produce paradoxes if applied to change or motion. But, outside of the discrete procedures of analytic thought there is an understanding which views time more the way it is consciously experienced as a continuum. The motion of objects needs a *flow* in time; their change of state requires a *direction* in time. Both of these need to be relative to a conscious observer in relation to which the object moves and/or changes.

Unlike Heraclitus who could effortlessly see how Achilles would overtake the tortoise, many centuries later Western science sided with Zeno. In so doing it rejected the model of an indivisible motion of organic and inorganic processes for a more easily analysable model of discrete separable events. Nevertheless, Heraclitus' river, waves, music, the passing of time and most important our very own conscious existence would cease to function if they were not intrinsically continuous. They can't be switched off and on and they can't be split up into discrete parts

Greek thought was adapted but without its metaphysical baggage. The irony is that Greek idealism was converted into an extreme pragmatism. While the Greek idealists were prepared to accept that reality itself is a changeless world of forms, the West needed to find an interpretation of time's flow and direction, in which experimental science could blossom whilst still keeping their world-view 'objective', that is to say in no way dependent upon the perception of the subjective observer

The problem of direction was 'solved' by Isaac Newton in the late seventeenth century. For Newton, time proceeds independently without relation to anything external. Neither consciousness nor the objective world affect this. Therefore its passing has no yardstick against which to give it a 'direction'. All events were theoretically time asymmetric. Equations would work unaffected by time's direction. No proof was offered for this hypothesis but none was needed. It was accepted as a theory because it worked in practice.

But if the direction of time could be safely ignored what about its *flow*? In the external world this appears as motion and since the sixteenth century this had been a central concern of physics. How could its motion, more specifically a change in the rate of motion, whatever the direction be measured? Leibniz and Newton solved this dilemma in a pragmatic way with the development of calculus. This was a mighty step. It enabled everything, even continuous indivisibles to be expressed in mathematical language. From now on all motion could be used in formulae and thus the whole universe was quantifiable. The use of delta 'tending towards zero' and later the concept of 'limits' meant that Zeno's paradoxes were apparently solved.

Motion therefore did not need an *observer* because its direction was irrelevant (all classical equations work regardless of time's direction) and its flow could be explained and calculated without the need to assume a conscious observer of an object's speed or change of speed.

And so, for functional purposes, the problems of time were resolved. Applied science was triumphant: engineers could use their tables to calculate the effect on different stresses and speeds on structures or machines, which helped usher in the 'machine age'.

Calculus works superbly and is the basis of a whole new field of mathematics, 'mathematical analysis. This uses the infinite series by which a limit is approached, as a quantity tends towards zero, but never vanishes. Previously only uniform motion from two fixed points on a straight line could be calculated as distance over time. The exact motion at any specific point on the line however could not be and variable motion, even a standard uniform change in the rate of speed, could not be calculated at all.

To sum up, the differential calculus is a brilliant functional device for by-passing the essential indivisibility of time to enable rates of change to be calculated. It does not attempt to make sense of time but simply bypasses the problems of its nature, getting round its indivisibility by using a quantity approaching zero so that mathematical calculations are valid. Scientific analysis needs a start and an end – that is, limits or boundaries. Calculus provides this.

The Western functional approach is that what works is true. The result was that from "1700 onward the human imagination was convinced that nature could be fully understood as a series of differential equations, as an algorithmic compression..." (1993: Appleyard, 43). Functionality had become reality.

Nevertheless, the paradoxes of Zeno still apply. There is no smallest unit of time. G.J. Whitrow in his *The Natural Philosophy of Time* (probably the best of the scientific books on time) writes: "examination of Zeno's paradoxes has led to the conclusion that for motion to be possible the point-like instant must be regarded as a logical fiction. It follows that we can accept this concept only as a mathematical device which is employed simply as an aid to calculation." (Whitrow: 1961, 157)

He also quotes the philosopher Alfred Whitehead who, in his book *Modes of Thought* (Cambridge University Press 1938, 207), states: "There is no nature apart from transition, and there is no transition apart from temporal duration. This is why an instant of time, conceived as a primary simple fact, is nonsense". In calculus the discrete is masquerading as the continuous with startling practical results. Zeno's paradoxes are circumvented, in everyday language 'fudged'. To express this in information technology terms only what can be expressed algorithmically and reduced to binary code is computable. Therefore analogue processes have to be interpreted digitally.

But it was around time's *direction* rather than its flow that problems arose in the nineteenth century. As we have seen direction was apparently made irrelevant by Newton's time, which flowed independently of everything else and all equations were time symmetric. Everything was in theory reversible and repeatable.

Nevertheless, this approach which is diametrically opposed to the "common sense" view of time in which kettles steam as they boil but never 'unboil' putting the steam back into the kettle, became less and less tenable as Western science progressed. "Why do natural processes always run one way while the laws of physics say they could run equally well either way?" (1999: Barbour, 24) This question had been avoided by assuming processes were *theoretically* reversible. But, as science advanced in the nineteenth century flaws could no longer be brushed aside.

The first serious crack to appear, which could not be ignored, was the concept of entropy contained in the Second Law of Thermodynamics. The First law of Thermodynamics developed from Newton states that the amount of energy remains unchanged, though it may change its form dramatically. This is the law of the conservation of energy. The Second Law of Thermodynamics, however, states that the amount of *useful* energy is being constantly dissipated.

The heat generated by mechanical processes is a random movement of particles in the system and is never fully recoverable. As the process continues eventually all the useful energy will be lost, so no more change can occur. This process was first deduced by Sadi Carnot in 1824. In 1850 Rudolf Clausius formalised the law and coined the term 'entropy' to describe the result of the law which is that the amount of disorder in systems grows. This final state is often referred to as 'thermodynamic equilibrium'. On a macroscopic scale this implies that the universe as a whole is moving towards a final state of irreversible disorder in which no further change can occur.

Entropy is not just a theoretical notion, but is used in statistical mechanics to make probabilistic predictions about the behaviour of complex systems. It looked as if the directionality of time had to be re-instated.

At about the same time as entropy was elaborated the theory of evolution was developed. The directionality of time was also a prerequisite for this. Natural selection is an irreversible process. But in sharp contrast to entropy an increasing order appears to be manifested. There has also been much progress in molecular biology, despite its tendency to treat living beings as if they are very complex machines. But still on a micro level it cannot fully explain how an egg forms a living being; on the macro level it cannot explain how man evolved to become a being so different from his ancestors and the rest of living creatures. The processes underlying life's ability to organise itself remain elusive.

A third directional force concerns the universe. In previous centuries astronomy viewed its origin as an insoluble problem from a scientific perspective. However, in the late nineteenth century astronomical tools became increasingly powerful and in the twentieth century a new science, 'Cosmology' emerged. Properties of the universe began to be investigated, including its genesis and development.

The Newtonian atomistic model needed either replacement, revision or supplementing. The traditional billiard ball analogy where in theory the balls could be set up again and a game replicated no longer applied.

Against this background of increasing acceptance of the need for 'time's arrow' arose new scientific discoveries which, appeared to demand a return once again to the concept of a directionless time which was being discarded by post-Newtonian science. The following is an overview of these discoveries. Its brevity and the focus on their impact on concepts of time inevitably involves some degree of oversimplification which, hopefully, will not be open to misinterpretation.

James Clerk Maxwell in the 1860s established mathematically that electromagnetic effects propagate at a rate equal to the speed of light. A few years later in the renowned Michelson-Morley experiment of 1887, rays of light were reflected back over long distances but no variation in their speed was detected. Lorentz had followed up Maxwell's work by developing a set of 'transformation equations' (equations of motion) demonstrating that electro-magnetic propagation, which includes light, can only be explained if the speed of light is absolutely constant, regardless of the relative position and velocity of light rays. Therefore two rays of light approaching each other on collision course will not meet at double their speed, as two trains travelling at the same speed would on impact. This is because the speed of light is an absolute limit, which cannot be exceeded. Electromagnetic theory produced brilliant practical results with the development of such devices which are now taken for granted, such as electric motors, dynamos and radio.

The late nineteenth century scientific breakthroughs were drawn together in 1905. Albert Einstein's Theory of Special Relativity created a brilliant new model of the universe. The constant speed of light was incorporated in his famous formula: e (energy) = m (mass) c^2.

Relativity, as the name indicates, repudiated old absolutes, most spectacularly in the case of time. In the Special Theory of Relativity the force of gravity is treated as negligible and is ignored. The effects of gravity were set out ten years later by Einstein in General Theory of Relativity ten years later in 1915 which he integrated the effect of gravity into relativity theory. Part of the effect on time of the Special Theory can be caused by motion alone. However, this only becomes apparent at extremely high speeds.

The famous paradox of the twins is a good way of graphically portraying the impact. If one twin Annie goes on a very fast voyage in space she would according to the Special Theory return younger than her stationary brother on earth. Annie in orbit travels with a clock. When she returns to earth the time shown by the clock of her twin brother who has remained on earth shows a later time than the time shown on the travelling twin's clock. The immediate objection to this is that time has not really gone slower for the clock or person only the instruments which measure it have been distorted by what is called a 'length contraction' effect. But the evidence contradicts this objection. Of course humans cannot yet travel at such speeds but unstable particles in the Hadron collider do: their short lived existence *is* observed to be longer the faster they travel.

The second effect being considered results from the Theory of General Relativity. This shows how gravity slows down time. So time for an object or person orbiting the Earth where there is less gravity goes faster. Objects on earth act as if they are 'weighed down' by the force of gravity. They have less energy than objects in space: this includes the clocks which measure time with the consequence that they run slower than they would in outer space. This effect is measurable for GPS satellites whose clocks have to be adjusted to synchronise with Earth's time.

These examples demonstrate how in relativity time is inextricably linked to space. The effect of gravity on objects will vary according to their *distance* from the massive object whose gravitational force is acting upon them and therefore how fast time runs for them. Compared to a person with his feet on earth there will be a faster time for someone in outer space or even from the top of a mountain or on the top of a tree.

Returning to the twin example above if the brother on earth had gone on half the distance in space with his sister in another spacecraft before turning back presumably his sister's younger age relative to him would only be half as great than if he had remained on earth. Their mother, when the twins have both returned to earth would find that the difference of twins' ages relative to each other was not the same as their difference relative to her. The puzzling consequence of this is that just as people cannot occupy the same space in relativity it seems they do not necessarily occupy the same time. All three people, the mother and the two twins, have shifted their perception on how old they are relative to each other. To explain how this occurs can only be done after a brief outline of time in relativity theory which is given below

Understanding time in relativity requires a completely new conceptual scheme far different from one Newton's in which time is a constant independent variable and far from our common-sense everyday notion of everyone being in one universal time which moves forward regardless of us. The relativistic treatment of time can be viewed in terms of the combination of a series of new discoveries and the theories developed from them:

1. Light had been found to travel through space at a constant speed.
2. In Einstein's equation $E=mc^2$ the speed of light is the constant element relating mass to energy. By this equation mass and energy are in a sense equivalent and one may be converted into the other. On a cosmic scale this does in fact occur. For instance the Sun loses mass as it is converted into energy.
3. In relativity an amount of time is defined by the distance light travels during that time. Time can now be measured spatially (confusingly 'light years' are used for convenient form of *spatial* measurement for vast distances).
4. With time being measured spatially it can be linked to space as another dimension compatible with the three spatial dimensions. In relativity events occur in this a

four-dimensional space-time universe. Minkowski, who taught Einstein, created a geometry based on these four dimensions and everything that applies to the three spatial dimensions can be applied to the fourth dimension.

Relativity has paradoxical effects. It leaves the old Newtonian-based universe unchanged for practical purposes as it only makes no appreciable difference to calculations on the scale of Earth which is a minute object in terms of the cosmos. Yet at the same time it introduces a breathtakingly different conceptual scheme. Time can no longer be a constant independent variable and is regarded as the fourth dimension. In this four-dimensional universe time is spatialised by defining its measurement in terms of a new absolute the constant speed of light.

The term 'time dilation' is used in relativity to refer to how it changes time. The above outline of relativity theory can now applied to the two examples of time-dilation described earlier. Starting with the twin paradox this illustrated the time-dilation effects of high speed travel on time. The discrepancy in time on their clocks reflects the difference in the increased duration of time that has passed for the stay-at-home twin. When the travelling twin meets his sister on his return their spatial position is now the same, but their positions in time have permanently changed. Their times are no longer simultaneous. This has nothing to do with the relative distances travelled. If I spent two years cycling round the world this wouldn't mean I'd feel younger compared to my wife when she met me on my return. Instead it results from the relative speed of the travellers. A good way to visualise relativistic effects is to exaggerate them. As a body or person travels closer to the speed of light their time slows down and their mass increases. (At the theoretical limit of travel at the speed of light time would stop and mass would become infinite.) On a much less dramatic level this conversion has affected the travelling twin. It is permanent and not reversed by the traveller returning to earth.

Turning to the time-dilation effect associated with gravity this can be summarised as follows:

- Gravity is what differentiates an object's *mass* from its *weight*. The amount of matter of an object is its mass which will not change with its location. Weight is the measure of the force of gravity on the object. This will vary according to the gravitational pull at the object's location

- Light is a form of energy. All forms of energy have weight (An empty box containing light weighs ever so slightly more than one that doesn't). Consequently light is subject to gravitational attraction. The nearer light approaches to earth the greater the gravitational pull and as the pull increases so light curves towards Earth.

- The shortest distance between two points is a straight line. The speed of light is constant and therefore a curved light will take longer to reach Earth. In relativity the speed of light is equivalent to time and thus time slows down.

It should be mentioned, unlike the effect of motion, gravity's time-dilation effect is incredibly minute. Its effect on objects orbiting the earth can only be detected by the highly accurate time measuring devices of modern technology.

In a relativistic universe we are in the apparently static four dimensional universe called space-time. Does relativity then require going back to go back to an explanation of time based on two types of time. One would be a perceptual time inside the brain, which is computational, flows and still has direction while out in the universe physicists have their time which is effectively a coordinate of space. This leads back to the concept of two times which was discussed and rejected in the earlier chapter on philosophy. Whether in physics and mathematics developments will force it to be reinstated can only be discussed after the effect on our understanding of time of quantum theory, the other great revolutionary development in modern physics, has been discussed:.

This book is about time and the above is an attempt to focus on gravity's effect on time rather than on space. This is a bit artificial because in relativity time and space can never be fully explained in isolation from each other. Nevertheless it can be helpful conceptually to do this. In relativity space, as well as time, is curved by gravity. In analogous fashion to my focus on time in the example above the curvature of *space* by gravity is often focused upon, for example, to explain the elliptical, rather than circular, orbit of the planets around the sun.

An examination of the combined effects of the curvature of space and time in relativity or quantum theory is not relevant enough to the subject to justify a lengthy digression. But even in limiting the discussion to the relationship between time and relativity some aspects have been omitted: for instance, the effects of acceleration and the rotation of the earth on time have not been mentioned. This is because the purpose of this section has been to the give examples of the way time is treated in relativity rather than to provide an exhaustive description of the subject

The New Physics and Objective Knowledge

Like Newtonian time was before, Einstein's time is now accepted because it works in practice. Parts of the consequences of his theory were only experimentally verified many years later a recent example being the existence of gravitational waves. The traditional Newtonian science including its understanding of time had achieved immense practical and theoretical triumphs, dominating the world of science for about three hundred years. Nevertheless, by the early twentieth century it was being torn apart. Firstly, as described above relativity theory overthrew the Newtonian concept of an absolute time by making time relative to the observer. Then Quantum Theory subverted both Newtonian and relativistic concepts of time as detailed later in this chapter.

From four-dimensional space-time the concept of a 'block universe' has been derived. As time has no directionality; there is no now or past or future; time is just another dimension of reality, independent of our consciousness. The block universe view "regards reality as a single entity of which time is an ingredient, rather than a changeable entity set *in* time "(Price: 1996,7). The block universe is a very powerful concept but not a "well-established scientific fact" as some claim (Ruckner: 1985: 149). The validity of the theory of relativity is in no way threatened if there are doubts about its status. Having reservations about the block universe theory is not the same as being a member of the Flat Earth society, rejecting the Copernican revolution and so still thinking that the Earth is the centre of the Solar System or refusing to believe in human evolution. It is a perfectly valid position to maintain that the block universe hypothesis is incorrect or at least incomplete. In fact, the block universe really only suits the convenience of a particular interpretation of the New Physics : "some physicists don't like the idea of a 'moving present', regarding it as a subjective phenomenon for which they find no house room in their equations. " Dawkins: 1998, 3)..

Earlier we considered the three factors which seemed to favour the introduction of temporal directionality into *classical* science. These were:

1. The history and origin of the universe
2. The directionality inherent in evolution and some chemical processes
3. The phenomenon of entropy

Expressing in these three factors more general and less scientific terms the questions which need to be answered are how could there be a beginning of the universe and how can some things within it get progressively more ordered whilst others appear to get progressively more disordered? The theories of relativity means that classical science's answers to these questions now need to be revised as detailed below.

1. *The question of an origin of the universe.*
Prior to the twentieth it was not known how large the universe is and this together with its origin (if any) were matters of pure conjecture. At the time of the publication of the General Theory of Relativity in 1915 knowledge of the universe was confined to the Milky Way. However, through startlingly swift improvements in telescopic range its vast scale became apparent very soon. By the mid-1920s it was established, through discoveries credited to Edwin Hubble, that there were many other galaxies, often much larger than the Milky Way. This discovery was followed at the end of the decade by evidence that they were all moving away from us. The universe is expanding.

The New Physics and Objective Knowledge

Einstein's relativity, with its static universe, appears superficially to be opposed to the concept of an expansion and he found it difficult to accept. Bur worse was to follow. Einstein's equations were developed by the Russian meteorologist Alexander Friedman to explain not just the expansion of the universe, but also its origin in what is now commonly called 'big bang theory' which began space and time. Before that there was nothing.

Much evidence supports the big bang and it also ties in with the overwhelming evidence for of an expanding universe. What is taken as conclusive proof came later in the 1960s when 'cosmic background radiation' was discovered which whose existence could only be attributed to a 'big bang'. This is caused by electro-magnetic subatomic quanta and an easily readable full description of this can be found in the book *The Afterglow of Creation* by Marcus Shown.

What happened before the big bang and how did it come out of nothing? Cosmologists use the notion of a 'singularity' to describe a point in space-time at which gravitational forces cause matter to have infinite density and infinitesimal volume. It is used in descriptions of 'black holes', when massive stars collapse from their own gravity. Many believe the big bang was a kind of singularity. How exactly this initial singularity worked is still debated but on one point there is agreement. The general laws of relativity could not apply for a singularity. Somehow the universe and with it time emerge out of this 'nothingness' which in the first few infinitesimally small fractions of a second is small enough to be subject to 'quantum' laws. Very shortly afterwards however the universe became isotropic, (looking the same from any direction), in both space and time. Any apparently directional forces are simply the after-effects of the initial big bang. This supposition enabled cosmologists to square the circle" Like calculus did in relation to the *flow* of time, cosmological theory based on relativity circumvents any problems to do with time's *directionality*.

2. *The directionality inherent in evolution and some chemical processes.*

So far our discussion of modern science has focused around relativity. Relativity has little relevance to organic tendencies towards self-organization. This third factor in favour of reinstating time's directionality can therefore only be looked at after non-relativistic scientific theories have been discussed later in this chapter. Nevertheless it is worth noting at this point that relativity theory, through its inability to explain the emergence of organization inherent in living organisms, cannot stand by itself as the basis of a 'theory of everything'.

3. *The Phenomenon of Entropy*

Is the directional force of entropy also a result of the big bang? There continues to be a great deal of speculation but no definitive answer, attempts to derive it from the big bang being fraught with contradiction. The fact is that in the symmetrical world of relativistic cosmology the all too evident asymmetrical nature of entropy is most uncomfortably situated. In isolated systems maximum entropy, known as 'thermodynamic equilibrium', will eventually be attained and no further change will occur. Ludwig Boltzmann demonstrated how, using probability theory, it was possible to describe how such relatively isolated systems would behave and when thermodynamic equilibrium would be reached. This was very useful for engineers and led to the development of a whole new discipline, statistical mechanics. He speculated that thermal disequilibrium is only a local phenomenon. The universe as a whole is in a thermal equilibrium, therefore directionless and compatible with the theory of relativity, the illusion of directionality being caused by our local viewpoint.

The French mathematician, Jules Henri Poincaré, is seen as the precursor of chaos theory, first known more appropriately as complexity theory. In his study of systems very far from thermodynamic equilibrium he showed that the techniques of statistical mechanics *cannot* be applied to them. These are systems where the interaction and feedback from their environment is continuous and often varied. Such systems are not uncommon. In fact most systems are not isolated, as common sense would tell us. They are the norm. The weather is the obvious example, but even very simple systems, if they are open, can become incredibly complex if they are not isolated. The effects of even slight variations in conditions can lead to totally unexpected results. They display an increasing disorder in accordance with the law of entropy, but are not statistically predictable. Could they be seen as proof that entropy is fundamental not just in isolated systems, but in the universe as whole? Universal entropy could not be disproved. In fact, the contrary occurred and in the second half of the twentieth century, with advanced computing techniques, its reality was established.

Besides relativity the twentieth century originated two other highly influential developments:

1.   Chaos Theory
2.   Quantum Mechanics

Ways in which these challenge the concepts time in the world of classical science as modified by relativity are now examined.

Chaos Theory

The full development of chaos theory quite a bit later in the twentieth century than quantum theory, which had become important in the first decades. Nevertheless, it makes sense to treat it first. There are many reasons for this, but the principal ones are:

- It relies heavily on the theory of entropy for its proposed arrow of time and is therefore closely linked to the preceding discussion of the relationship between entropy and relativity.

- This reliance on entropy is so strong that many proponents of chaos theory believe that its explanation of entropy can alone explain the arrow of time.

- In making this claim it implicitly, for some proponents explicitly, denies that quantum theory has a role in the reintroduction of time's directionality into physical science.

Chaos theory itself emerged only in the 1950s and 1960s when computers developed the power needed to analyse the complexity of the behaviour of systems far from thermal equilibrium. The results were surprising. Some systems did indeed eventually return to their initial state. But other systems, often quite simple to begin with, become more 'chaotic'. These systems exhibit entropy as they grow increasingly disordered but never reach maximum entropy. Their change though, is irreversible. As irreversible change is how we can be aware of time's directionality and, as chaos theory shows how that irreversible change occurs, it is, chaos theorists claim, chaos which gives the arrow to time.

However, this is just brushing the surface of chaos. Just as relativity is a misleading name as it introduces a new absolute – light, chaos theory is misleading as it introduces a new type of order! In these systems far from equilibrium even the smallest fluctuations can lead to huge changes. Out of chaos order can and does emerge, which becomes apparent as computers crunch through vast iterations of the systems with the feedback that causes the complexity. Patterns appear in these systems, whether it be geographical formations, population trends, stock market prices or electrical circuits.

The Nobel Prize winner, Ilya Prigogine, whose specialism is irreversible chemical thermodynamics, takes this a step further. Not only are there irreversible processes but these "are the really fundamental entities in the universe and that the idea of microscopic particles moving subject to reversible laws is an approximation that is only valid where a particle, or particles in co-operation, are effectively decoupled from their interaction with the rest of the universe" (Rae: 1994, 108).

Sometimes these can lead to what Prigogine terms the emergence of 'self organization'. This self-organisation is apparent in the phenomenon of life where extremely complex biological processes result in sophisticated ordered beings, such as ourselves. But it does not just apply to living beings. Prigogine demonstrates how self-organisation in chemical processes far from thermal equilibrium is maintained. They create order from disorder by a process which could be loosely described as adding new chemicals to a mixture. In Chaos theory against the backdrop of overall increasing entropy, due to self-organisation there is a great deal of local directional evolution taking place.

Understanding Time

James Lovelock in his Gaia theory applies this self-organisation to the Earth as a whole. Organised life forces help maintain the earth's environment by feedback between them.

While chaos theory claims that the necessity for a directional time arises from physical properties of the external world quantum theory was leading to an even more radical basis for the directionality, indeed for the very nature, of time.

2. Quantum theory.

Quantum theory challenges not just Newton's absolute and relativity's spacetime. Treatment of the flow and directionality and indivisibility of time is entirely differently from classical science. It also brings into question the underlying assumption of Western science that time is independent of consciousness.

Like relativity, the development of quantum theory was closely linked with investigation of electro-magnetic phenomenon, in particular light. In Maxwell's electromagnetic theories light was assumed to have a wave-like nature, which was confirmed by 'interference' effects. These showed that if two emissions of light met they could reinforce each other or impede each other depending on their relative frequency, just like waves in the sea. However, around the end of the nineteenth century, experiments with sub-atomic particles, notably electrons, produced results that showed light behaving more like particles, conserving energy and bouncing off electrons. Subsequent developments confirmed that light exhibited this dual wave-particle nature.

These wave-particles were visualised as wave-packets and given the name 'quanta'. In the 1920s Werner Heisenberg and Erwin Schrodinger formulated the principles of what became known as quantum mechanics. Schrodinger developed probabilistic equations which could predict probabilistically the development of the wave packets. These equations were totally time symmetric and deterministic so did not clash with the tenets of classical physics.

Another principle that came out of this wave-particle duality was less so: Heisenberg's uncertainty principle. You can either measure the wave like nature or so discover wave-like properties like its speed, or you can measure its particle-like nature and determine particle-like properties like its position. But you can't do both at the same time, so either its position or its speed will be undeterminable.

Even more extraordinary were the consequences of taking a measurement. The probabilistic nature of the wave then 'collapses', with the particle being found at one position only. This collapse is irreversible and so an 'arrow of time' appears. According to some followers of what is referred to as the Copenhagen Interpretation role this collapse only occurs for a conscious observer.

While relativity introduced the notion that time is relative to the observer, that observer might as well be a CCTV camera:

"Einstein's time, despite its limited *observer-dependence*, still adheres to Laplace's determinism, to a rigid chain-mesh of cause and effect, in which the destiny of the world has been etched into the fabric of nature since the dawn of existence." (Davies: 1995, 274 – my italics)

Einstein's universe often imagined as one in which a nothing has a history or a direction. Everything just occupies a position of the world line and which is theoretically reversible. A 'quantum collapse' involving consciousness as an active causal participant threatens this.

Understanding Time

The ontology behind the 'quantum collapse' is still a matter of intense debate. Roger Penrose devotes a complete chapter (Penrose: 2004, 782-815) to these, listing six alternative explanations. Many of these are desperate attempts to refuse to accept implications of the Copenhagen interpretation. As this debate directly concerns the central issues of this chapter, indeed of this book, what the 'conscious observer' signifies is now expanded upon.

The notion that an event depends upon conscious observation leads to what appear to be impossible paradoxes. There is the famous one of Schrodinger's cat. The cat is in a box with a radioactive material which has a 50-50 chance of decaying in the next hour. If it does it will trigger the opening of a poisonous tube of gas which will kill the cat. Will the cat be dead at the end of an hour? If its death depends on conscious observation we have the absurd situation that someone has to open the box before the cat either dies or survives, being in an indeterminate state till that happens.

There are many ways round this paradox one notable example being 'Wigner's friend'. This character, with a gas mask on, stays in the box with the cat. Wigner's friend can be asked what happened before Wigner opens the box. The friend's consciousness ensures that no quantum collapse is necessary, whereas the presence of just a measuring device would not. The situation however is by no means resolved. Supposing Wigner's friend fell asleep or, as Einstein suggested, that the observer was a mouse who observed but could not be interrogated by the human who opens the box.

Rather than attempting to resolve this by individual conundrums like Schrodinger's cat a much more fruitful approach is to examine what is meant by consciousness, particularly in relation to the role of consciousness in quantum collapse. This is the crux of the debate between idealist and realist interpretations of the new physics.

On the one side some continue to deny any role for consciousness: "The far-fetched nature of both the consciousness-based and many worlds interpretation of quantum mechanics is typical of the difficulties posed by the measurement problem within the existing framework of quantum theory. The fundamental problem is obscured owing to the widespread yet fallacious belief that the conventional Copenhagen interpretation requires the existence of 'observers' – assumed to be human beings – thus opening the door for subjective elements to creep in. But there is no need for conscious beings to be present: it is enough for there to be a measuring apparatus. For example, it might consist of a computer, a phosphorescent screen, a photographic emulsion or a bubble chamber. Human minds are irrelevant. Wave function collapse is irreversible and fully objective" (Coveney & Highfield: 1990, 134)

This goes hand in hand with a strong realist approach as another quotation from the same book indicates: "With all the evidence amassed by science in favour of an independent reality out there it is hard to take the consciousness-based approach to quantum theory very seriously. *Indeed, it may one day even be possible to understand consciousness in physical terms* (Coveney & Highfield: 1990, 132 my italics).

The end of this quotation points to one of the long term aims of realist thinking – to reduce mind to an epiphenomenon.

Idealists interpretations have built on the Copenhagen interpretation to solve the quantum measurement paradox in an entirely opposed fashion. Rather than examining the role of an individual consciousness they posit consciousness as a general phenomenon, which is actualised in the minds of conscious beings. External reality only exists in potentia. "The transition of possibility to actuality, or the collapse of the possibility wave to the actual particle, is an act coming from freedom of choice of our consciousness" (Goswami: 1998, 2). The form the collapse will take will depend on the form of that consciousness, thus getting around the difficulty of a mouse or cat's view.

Both approaches can never be proved empirically because it is just not possible to know the exact trajectory of a wave-particle between observations.

Before going further into the theoretical ramifications of time in quantum theory at this point a brief digression is taken. Returning to the notion of physics as a 'dappled world discussed at the beginning of this chapter it is worth attempting to reconcile these opposing views by limiting their applicability to separate realms each of which is relatively unaffected by the other. Both may be true, at least in functional terms, if their relevance is thus restricted.

Our use of time could be divided into two realms time as observed (by its measurement) and time as experienced (by its transience).

The ambiguous status of light could be viewed in this way. On the one hand it is susceptible to treatment in an atomistic fashion as a series of particles or points makes it compatible with objective time. On the other hand light's wave properties makes it more resemble mental 'subjective' time with properties of flow and direction.

A river is continuous and irreversible when you experience its flow, but can be treated discretely when you *measure* its rate of flow by using calculus. By analogy human actions are continuous when you experience them by doing something, but can be discrete when you observe them. The act of conscious observation instigates the transfer from the immanent flow of the internal world to the measured motion of the external world. It collapses the wave function into a defined position.

This 'dappled world helps in terms of clarity and functionality so that the conflicts between relativity and quantum theory and the more general conflict between science and 'common sense' don't interfere with our everyday life which we can just get on with. It seemed a good point to digress to point this out before going more deeply into the argument. The remainder of this chapter is not absolutely essential to the point of view being developed and can safely be skipped if it is not of interest.

The idea that in reality we live in this 'dappled' universe which runs on two types of time is very unsatisfactory. Can there really be two times, an objective 'real' time out there in which the laws of nature apply, and a subjective 'ideal' time and how do these two times relate to each other? At this stage it should be apparent that I regard objective time as less fundamental than 'internal time'. It is the spatial reflection of time. There must be conscious sense of time which reacts with externality in general prior to the observation of an external world. As quantum physics demonstrates any absolute division between an internal world and external world cannot be maintained. They interact as in the case of a quantum collapse. "The classical ideal of an objective description of nature is no longer valid. The Cartesian partition between I and the world, between the observer and the observed, cannot be made when dealing with atomic matter)" (Capra: 1976, 79).

The effect of quantum theory and chaos theory on the whole of the physical sciences is already far-reaching. With their reintroduction of direction the status of the algorithmic techniques to measure change are reduced to what they always were, very convenient functional devices. The timeline doesn't just measure the distance between different points in time but the passage from past to future. A point on the line, however much it tends towards zero, no longer exists by the time of the next point. There can be no precisely defined causal chain. The Heisenberg uncertainty principle means that at least one variable cannot be predefined:

"In quantum mechanics we find that the past history of a system does not determine its future in any absolute way, but merely the probability distribution of possible futures. In general, there is no conceivable set of observations that can provide enough information about the past of a system to give us complete information as to its future. (Whitrow: 1975, 294)

The foundations of the elaborate edifice of modern science are becoming shaky. It is dismantling itself rather than being toppled by outside forces. Recently it has become acceptable in the scientific community to argue against the cornerstones of classical physics. Time asymmetry, the irrelevance of consciousness, the physical reality of mathematics and the materialist basis of knowledge are all being questioned.

The revised paradigms of Time are very much part of this. Both chaos theory and quantum mechanics agree on the primacy of becoming over the become, which can be interpreted as the primacy of time over space. According to Chaos theorist Ilya Prigogine the time-symmetric block universe of relativity is a restricted view which works admirably, but does not apply to the more complex chaotic systems which can emerge very easily out of very simple initial conditions.

"A more general theory embracing both quantum mechanics and general relativity may be intrinsically time-asymmetric... we may be on the threshold of a radically new framework in which time occupies a central rather than a marginal role." (1990: Coveney & Highfield, 296).

Quantum mechanics also reveals a world in a perpetual state of becoming with 'tendencies to exist' only being actualised as the 'become', when they are made immanent by consciousness.

The becoming 'continuum' approach has always been the mainstream view in many Eastern traditions. Actually, it was never really smothered in the West. The effect of the reinstatement of the conscious observer in relativistic concepts and consciousness in general in quantum mechanics has provoked interest in the non-materialist views of different advanced civilisations. Fritjof Capra's book *The Tao of Physics* stresses the resemblances, between Chinese and Indian religious and metaphysical concepts and those used in modern physics. Another revival is Stoicism. Studies on the Stoics, who fundamentally opposed an atomistic approach to time, never really ceased, particularly in France. Ernest Brehier and Victor Goldschmidt are two twentieth century academics who specialised in the Stoics. The latter devoted an entire book to the study of the Stoic idea of time. As one of the phenomena which the Stoics referred to as 'incorporeals' time had no independent objective reality. For the Stoic, time is *qualitative* not *mathematical*.

Against this backdrop science is still striving towards a theory of everything which can embraces and reconciles quantum physics and relativity. The problem is that these two are not consistent with each other:

Understanding Time

"It has been the strange destiny of twentieth century science to have evolved two theories of immense power – some say the most powerful ever – only to discover that they appear to be contradictory." (Appleyard: 1994, 147)

At present this theoretical inconsistency may be 'liveable with'. Nevertheless, it is beginning to have an influence. This view is supported by Roger Penrose who refers to "a direct clash between the foundational principles of quantum mechanics and those of general relativity" (Penrose: 2004, 849).

The implications are already strong enough for conflicting world-views to be voiced. A materialist/reductionist camp and idealist/holistic camp have been setting up their tents. This is not such a new development. Many of the most distinguished scientists have been outspoken about the metaphysical, epistemological and sometimes mystical implications of their work. Einstein's comment 'God does not play dice' expresses his misgivings about the indeterministic outcomes of quantum theory. Niels Bohr was aware of the parallels with Eastern religion of quantum theory. Then there is the interesting case of David Bohm, very much admired by Einstein and seen by him as a likely candidate for bridging the quantum theory/relativity divide and whose work is described in more detail later in this chapter.

Chaos theory is a newcomer and its impact on the scientific perceptions of time is difficult to assess. However, in general, it has presented itself as firmly materialistic. A great deal of attention has been paid to making it consistent with relativity and Newtonian mechanics. Yet at the same time it is not reductionist in the same way, pointing to processes, especially organic living processes, which are far too subtle to be captured by a universal determinism and which could not occur in a strictly time-symmetric model.

The New Physics and Objective Knowledge

For the materialist often the most indigestible part of acceptance of a role for consciousness is that this admits unreliable subjectivity into the realm of science. But this is simply not true. Distinguished neurologists have worked with definitions of consciousness which by their universality can be strictly applied and tested:

"Subjective entities require, as do objective ones, that enough observers undertake rigorous observations according to the same experimental design; and they require that those observations are checked for consistency across observers and that they yield some form of measurement...knowledge gathered from subjective observations.. can inspire objective experiments, and, no less importantly, subjective experiences can be explained in terms of available scientific knowledge" (Damasio: 2000, 309)

With the admission of consciousness the floodgates open and in pour all the currents, which the gates of materialism tried to hold at bay. Mind is no longer separated from matter but mingles with it and it becomes problematical to distinguish the two. This threatens the materialistic view of time. Indeterminism, time asymmetry, irreversibility, becoming's primacy over the become are all plausible if time is partly or primarily a property of consciousness, rather than of the physical world.

Against these idealist trends there has been a materialist backlash. Two prominent examples are in interpretations of quantum collapse and research into the nature of consciousness.

Since the 1950s a theory of quantum collapse now called 'the many-worlds' hypothesis' has attracted a lot of support. Instead of collapsing the wave function into one reality, all possible outcomes are actualised. The wave function does not collapse but goes on evolving different, mutually unobservable, but real worlds. This (at a very high cost) saves the symmetry of time, the physical reality of mathematical entities and does away with consciousness playing an *active* role.

Another hypothesis is raised in the theory of *decoherence*. Like the many-worlds hypothesis the wave does not collapse when measured. Instead its superpositions 'decohere' into particle paths after contact with matter. The Copenhagen Interpretation would only apply in the extremely isolated circumstances of a controlled experiment. In practice, the world is much more open. The surrounding environment gets 'entangled' with the quantum wave-particle which 'declares itself' in a definite location. A conscious observer is not necessary as the 'declaration', even if not observed, would happen spontaneously in interaction with the surrounding environment. Many hold that decoherence has finally put the cork back in the bottle of mind to stop it overflowing into material reality..

The commonly held scientific view that materialism is the only serious option is also illustrated by Daniel Dennett's contributions. Dualism must be rejected outright because there is no place in the brain that can act as an interface between mind and matter. This means that time itself must be material. Dennett can then state "the brain's representing of time is anchored to time itself." (Dennett, 199,151)

Another current battleground involving the role of consciousness is in neurology. How does spiritual or religious experience arise? Doctor Persinger exemplifies the diehard materialist position that all such experience originates in parts of the brain and that using the experimental method is the only valid way to distinguish truth from non-truth. Apparently epileptics are more likely to have spiritual experiences because of abnormalities to their temporal lobes. More generally, the loss of self and changes to feelings of space and time occur in our parietal lobes. So a new branch of neurology called neuro-theology attempts to examine and initiate spiritual experiences by monitoring or stimulating the areas of the brain with which it is associated. (Horizon, BBC2 'God on the Brain' April 17, 2003) However, all this proves is that spiritual experiences are *associated* with certain areas of the brain. These may be *activated* in meditation or the attachment of electrodes, but the truth of the experiences are neither validated nor invalidated. The sensation of time during these mental states is that one in which it is free of attachment to the world as phenomenon. It is not surprising that this is reflected in unusual types of brain activity. But, observing correlations between mental states and brain states proves nothing of the truth or falsehood of the mental state.

However, with dualism discredited materialism is not the only option left. There is still a choice between (a) all is mind and (b) all is matter, technically called idealist and materialist monism respectively. .

How does the mental phenomenon of belief relate to the accompanying physical manifestations in the brain? You can induce changes in mind state through the brain or you can create changes in brain state through the mind. The question remains open whether the brain is an epiphenomenon of consciousness, or consciousness an epiphenomenon of the brain or that they interact through quantum signals. The last option is discussed later in this chapter. From a monistic idealist position what is important is the agreement of the mind with itself. This can be experienced in a non-dualistic feeling of harmony and oneness which can be triggered in the mind during different states of consciousness

Understanding Time

Once again, these mind-matter disputes become clearer when examined in terms of the role of time. Time is evidently only sensed by conscious beings and cannot be found by analysis of any material thing. It may well be possible to provoke a vague religious-like feeling by stimulating parts of the brain. But reducing a revelation in which the whole truth is revealed, whether it be Newton's apocryphal apple or St Paul's conversion on the road to Damascus, as relating to a particular brain state is far-fetched. As with our approach to time in discussing freewill and creation for everyday events objective time is a useful approximation, because for most events, just as with most thoughts and acts, the impact of internal time is so minimal it can be ignored. But at the limits, objective time is transcended in neurology or physics, just as it is in great art or great actions. The internal time element can no longer be treated as non-existent. All parts of the event are inseparable spatio-temporally and no measurable causal chain in linear time can explain them.

The idealist/materialist rift also extends into cosmology. The results of using quantum theory in connection with time's role in the origin of the universe can lead to diametrically opposed views, which appear to depend more on the scientists' overall *weltanschauung* rather than scientific evidence. The 1983 Hartle-Hawking theory uses abstract mathematical concepts derived from quantum theory and relativity to arrive at a tentative explanation. According to this, at the very beginning in the first nanoseconds there were four space dimensions. There was no first instant of time, just a change in the nature of one of the four space dimensions which became temporal. Time emerges out of space.

The time treated here is a mathematical convenience. Time reduced to its role in time-space is not real and lived in. In logic and mathematical functions it works but can provide no insight on the nature of time. These speculations are proposed in living time and, although they are made by scientists, they are as far removed from objective scientific proof as the ideas of artists, philosophers or the 'man on the Clapham Omnibus'. Time is qualitative and not reducible to quantities. To treat time like a quantity is to be like the money-obsessed who "know the price of everything but the value of nothing". As a quality it has taste, purpose, direction - in a word all the non-algorithmic properties, which computers cannot have and never will be able to have.

A scientist at the other extreme from this is Amit Goswami:

"The truly cosmological questions can now be answered: How has the cosmos existed for the past fifteen billion years if for the bulk of that time there were no conscious observers to do any collapsing of wave functions? Very simple. The cosmos never appeared in concrete form...the universe exists as a formless potentia in myriad possible branches in the transcendent domain and become manifest only when observed by conscious beings" (Goswami: 1993, 140-141).

In Amit Goswami's view, quantum theory "is paving the road for an idealist science in which consciousness comes first and matter pales to secondary importance" (Goswami: 1993, 62). The corollary of this is that time does not emerge from space, rather the reverse:. "...the universe exists as formless potential in myriad possible branches in the transcendent domain and becomes manifest only when observed by conscious beings" (Goswami: 1993, 141).

Goswami here is using quantum theory to propose that there is no permanent world extended in space but an external reality whose very existence and nature depends on a quantum 'collapse' which only happens if consciousness intervenes:

"In the idealist interpretation, time is a two-way street in the transcendent domain. When consciousness collapses the wave-function of the brain mind, it manifests the subjective one-way time that we observe. Irreversibility and time's arrow enter nature in the process of collapse itself, in quantum measurement." (Goswami: 1993, 102)

If consciousness is the ground of experience, as idealism proposes, it is a meaningless question to ask what state something is in without consciousness.

Paul Davies in an overview of quantum time states this consequence of quantum theory succinctly. "The common sense idea that there is an objective reality "out there all the time" is a fallacy. When reality and knowledge are entangled, the question of *when* something becomes real cannot be answered in a straightforward manner." (Davies: 1995, 173)

Many are feeling the need for some framework which can synthesise materialism and idealism. Otherwise we risk having two different world-views, rather than two different insights on the same world-view. The whole issue will not go away. 'When something becomes real' is part of the quandary of quantum mechanics and it is around another controversial aspect of quantum theory – the concept of non-locality, that new ideas and approaches are emerging.

In non-locality two connected events appear to happen simultaneously however far apart they are situated. This arose in quantum theory when it was proposed that if an electron changes its 'spin'a correlated electron must instantaneously change its 'spin' direction too, regardless of the distance involved. This was too much for Einstein who refused to accept implications of action at a distance, apparently involving superliminal speed. Nevertheless in 1982 a French team of scientists led by Alain Aspect proved that non-local correlation is a reality. If the wave function of one quantum object is collapsed (by measurement) its polarised spin-correlated partner will also collapse simultaneously, adjusting its spin to complement its 'partner'.

Parallels were made very early on with mental phenomenon, in particular with Jung's concept of 'synchronicity', where two events are linked causally and much too improbably for the link to be put down to coincidence. Freud was horrified by Jung's interest in synchronicity when the latter started connecting it with paranormal phenomena and astrology, expressing his concerns that Jung was no longer being strictly scientific. Yet Jung thought he was being scientific seeing connections between his methods and those of the new physics. "As soon as a psychic content crosses the threshold of consciousness, the synchronistic marginal phenomena disappear, time and space resume their accustomed sway, and consciousness is once more isolated in its subjectivity. We have here one of those instances which can best be understood in terms of the physicist's idea of 'complementarity'.

The physical side of the complementary relationship can be expressed as follows. It rests with the free choice of the experimenter (or observer) to decide which insights he will gain and which he will lose. It does *not* rest with him, however, to gain only insights and not lose any. "Physics determines quantities and their relation to one another; psychology determines qualities without being able to measure quantities. Despite that, both sciences arrive at ideas which come significantly close to one another... Between physics and psychology there is in fact 'a genuine and authentic relationship of complementarity". (Jung: 2001, para 440)

Interpretations of non-locality move outside the parameters of Newtonian physics and relativity. They rely on a non-spatial linkage where somehow the two events are linked because they are part of a whole which is more fundamental than any spatial separation. It is not surprising that against this background suggestions have been made to search for a deeper level at which unity can be achieved. This has led to a stepping outside of relativity and quantum mechanics into more all-embracing structures, not always traditionally regarded as scientific. Non-locality has been extended beyond quantum physics to the whole relationship between consciousness and the world.

The need for an integrating holistic approach runs through the development of quantum theory. Professor Pauli developed his complementarity approach. Niels Bohr stressed the notion of quantum wholeness in which there can be no fundamental separation between the observed, the mechanism used in the observing and the observer. Bohm took this to its logical conclusion in his 'implicate order. For him there is a level of reality that exists below the quantum and the normal macroscopic world and links them. He believed at this underlying level matter and consciousness are united in an underlying whole.

Another symptom of this holistic tendency is the removal of barriers between science and other disciplines. F David Peat, a close colleague of David Bohm, has pointed out how the inseparability of non-local phenomena is valid in many areas outside of science. He draws attention to the ways in which the strict local separation of the 'local' world cannot capture the non-local treatment of space and time in art and music, nor with many mental phenomena, in particular memory. Parallels with psychology are frequent, once again Jungian approaches being common, especially the role of a 'collective unconscious'. This has been portrayed as a psychological equivalent of Bohm's implicate order.

These holistic approaches are often associated with two strands of philosophical thinking – traditional idealism and phenomenology. An idealist viewpoint in which mind is viewed as real and matter as secondary becomes acceptable in the scientific realms. Amit Goswami has supported an idealism with roots going back to the Indian Vedas. His book *The Self-Aware Universe* is subtitled 'How consciousness creates the material world'. Henry Stapp, an American quantum physicist who worked with Professor Pauli, is another contemporary idealist. His view approaches Bishop Berkeley's idealist message '*esse est percipi*' (to be is to be perceived).

Phenomenology can be either realist or idealist. Phenomena are simply the objects we perceive. This is all we can know and the aim of phenomenology is to describe these as clearly as possible, without letting any preconceptions distort our perception. For example, any notion of an unknown reality lying behind them must be ignored. Husserl is generally acknowledged as the founder of modern phenomenology and one of the first users of the keyword 'intentionality'. Intentionality is the characteristic of consciousness to be always directed at an object outside itself. All phenomenon are perceived with this intentionality. How much this conscious intentionality affects the nature of the object is a moot point. Husserl tended to idealism, stressing the influence of consciousness whereas other phenomenologists, particularly those of approaching it from existentialist philosophy stressed the interrelatedness of different forms of appearance of the same phenomenon. The appearance might vary across time or cultures. It therefore must have some kind of being, apart from consciousness. It is a broad school, including Merleau-Ponty and Heidegger, both referred to in earlier chapters. But the connection with quantum theory lies in how the intentionality of consciousness can be likened directly to the way the choice made by the observer affects what he sees in a quantum collapse. Phenomenology blurs the boundaries between what pertains to conscious perception and what pertains to the objects of that perception in a manner very similar to quantum physics.

It did not take long for eminent quantum physicists to link all this together to develop quantum theories of mind. The choice made by a conscious mind in quantum theory became seen as an integral part of the reality. As early as the 1930s John von Neumann and Eugene Wigner were developing a quantum theory of mind in which consciousness interacts with our brains with non-local effects. Their theory became more testable and was boosted greatly by the Aspect experiment. Quanta could be seen as a product of the mind before the collapse in which the separation of the consciousness and the external world is accomplished. Henry Stapp takes an extreme view where each act of observation changes the state of the universe, the universe being a vast collection of observations changing its state continuously with each wave function collapse caused by observation.

Quantum non-locality has provoked a veritable explosion of theories of consciousness. One example is 'The Non-Local Universe: The New Physics and Matters of Mind' (1996: Nadeau and Kafatos). For these authors, quantum non-locality explains how consciousness came out of the brain's physiology and how the universe is conscious. The questions about time risk getting hopelessly entangled with these issues.

A masterly quantum interpretation of mind is in Danah Zohar's book 'The Quantum Self'. This provides a convincing quantum mechanical model of consciousness. The complex sub-atomic nature of neuronal processes in the brain is exploited to provide a plausible scientific scenario for quantum interfaces between 'mind' and 'matter'. A new non-dualistic 'holistic' paradigm is described which may become acceptable in time even to the most diehard materialists.

As with the issue of the quantum collapse earlier in this chapter the role of time will now be looked at directly using the technique of dividing the notion of time into two realms – measured time and experienced time. Using this technique a direct parallel between a relativistic and quantum-type interpretation can made. In the relativistic interpretation we have measured time. The external world time is part of the spatio-temporal scenario (space-time). In space and external time a *position* is only given relative to an observer (or a specific point of observation). In the external world the only absolute not relative to the observer is the speed of light as the underlying medium of external reality. The individual observation presents outer spatio-temporal reality relative to the observer, but *speed* of light itself is not relativized by the individual observation.

In total contrast, the *inner* world is intuited through time as the underlying medium of internal reality, but the *flow* of time is not relativized. The directionless *speed* of light in the external world is analogous to time's directionless *flow* (flux), in consciousness. In the internal realm a *direction* is only given when consciousness is related to a specific mind, i.e. a self-consciousness.

Understanding Time

Links between entities in consciousness are through their inseparability not by an external spatio-temporal causal chain. This is how the world appears for a consciousness before a focussed intuition, an *intentionality*. Time is localized in the *now* of a conscious being while space-time is localised in the *here* by an observing being. A non-local connection is where the mind unites phenomena inseparably in such a way that it overrides any external localisation. As in the earlier discussion on quantum collapse, 'mental' time is seen as more fundamental that 'physical' time.

What lies under the localisation in the *now* is beyond the limits of our individual consciousness, though maybe not of a collective unconscious. This is close to the view of the phenomenological philosopher Heidegger, a student of Husserl's who considers that "time is the possible horizon of any understanding of being" (Heidegger: 1962, 1).

"Science is mobile, its nature is constant change. One generation's certainty is likely to be overthrown by the next. It may be true that quantum mechanics points to a deeper spiritual realm – but knowledge of that realm must come from outside the quantum, otherwise it remains dependent upon the whims of science. We must, in effect know the truth before we can discover it in the quantum." (Appleyard: 1994, 216)

As has been shown, attraction to the philosophical, holistic or even mystical implications of quantum theory is quite common among leading quantum physicists. Perhaps they feel that the external evidence of quantum research would rest on more secure ground when supported by insights not just from other scientific disciplines, such as neurology, but also from non-scientific fields. The cross-fertilisation between quantum theory and the arts, philosophy and mysticism has already been immense and the benefits have been reciprocal. Other insights can be more widely accepted by a requirement to be compatible with, or complementary to, the physical sciences, in particular with quantum theory. Quantum theory and science as a whole benefit by becoming less dogmatic and divorced from other areas of human knowledge and expertise.

The next chapter looks at time in terms of a fusion of all the perspectives covered in this and the preceding chapters.

Chapter Six

What is Time?

Do you believe in future everlasting life?"
"No, not in a future everlasting life but in an everlasting life here. There are moments, you reach moments, and time comes to a sudden stop, and it will become eternal"
"You hope to reach such a moment?"
"That's hardly possible in our time", Stravogin said, also without the slightest irony, slowly, as though pensively "In the Revelation the angel swears that there will be no more time."
"I know. That's very true. Clear and precise. When all mankind achieves happiness, there will be no more time for there won't be any need for it. A very true thought."
"Where will it be hidden?"
"It will not be hidden anywhere. Time is not an object, but an idea. It will be extinguished in the mind"

Dostoevsky, *The Devils*: conversation between Kirov and Stravogin. Part Two, Night. Section 5

Perception.
Despite the title of this chapter because of the inscrutability of time it is as much about what time is not as about what time is. From our nature as human beings we know that our *perception* of time is species-specific. It is not the perception of an absolute time and has no objective priority over time as perceived by other sentient creatures. A claim of 'superiority' in thought over other creatures may be justified, but the same cannot be said for perception. Different species have varying degrees of acuity in hearing, sight, touch or smell, often obviously more powerful

than that of humans. Indeed, it seems that as our intelligence has developed much less of our innate perceptual capability, apart from short distance seeing, is utilised. It is possible, however, to recover it, as witnessed by the increased auditory and tactile responsiveness of visually impaired people.

Conception.

As for different *concepts* of time none of these has any more objective truth than another. However, in this case the differences are apparent by comparisons of various cultures and backgrounds, rather than contrasting humans with other species whose conceptual abilities we cannot be aware of. Our awareness of time, therefore, is influenced by both our physical organs of perception and our faculty of conception (which is in turn partly conditioned by our upbringing). But there are important differences in how these two factors affect our awareness of time.

Consider first the influences on our concepts. At one extreme take an almost totally conditioned person. He or she doesn't reflect much (so very little critical insight or understanding), never does anything independent (so all actions are more or less predictable) and certainly has little imagination with no creative impulse for free expression. Such a

person would, for example, uncritically accept the 'objective' view of time of modern science as it filters through to her or him through the popular press and television as 'the truth'.

At the other extreme take someone whose thought and actions are more original and less hidebound. A common characteristic of instigators of change, artists or scientists is that they transcend their conditioning. From this perspective, the insights of great thinkers, artists, mystics and lovers are just as valid as the findings of great scientific innovators, even though their terminology or means of expression may be equally obscure to some. As a cultural veneer is stripped, we can begin to view more clearly a less conditioned time. Nevertheless this can never be a disembodied time totally independent of human sensory consciousness. But, although we can't 'get there' we can approach 'it' , receiving intimations of it in art or love as well as using it practically in free action or when applying the abstract findings of mathematics and physics to the world around us.

Developing our time sense, to a much greater extent than with our space sense, happens when we change our mental viewpoint. As we alter our consciousness there are different accompanying shifts of time sense.

Along a sliding scale human beings are in the odd position that they can develop their time sense and approach eternity, but never totally arrive there. To know unlimited time conceptually we would have to step outside our own minds. However, within our limited understanding we can hypothesise entities that may actually be in an absolute or much less conditioned time. As examples:

There may be entities for whom time is completely external because they are eternal and can see time as something which mortal beings exist in. There may be some supreme force which creates time and so time is internal to it. Lastly there could be a consciousness that sees our time as a sort of space in that all the events of the world are seen as simultaneous as if they are for our human perception in space.

Time is not just an inner sense enabling us to conceive events as successive, but an outer sense too by which we order our perceptions. While our *concept* of time is shaped by influences on our mental development our *perception* of time is formed by the properties of our sensory organs. But just as our mental state can be developed our perceptual abilities may be able to be stretched. We know that we can increase our *spatial* perception either looking through a telescope or microscope, or changing our vantage point such as climbing a hill or boarding an aeroplane. But are there methods of increasing our *temporal* perceptions?

In Ouspensky's analogy of time "We see the world as through a narrow slit" (Ouspensky: 1981, 33). Is it possible to enlarge this split called the present by extending our awareness beyond our normal range of perception? One of the main theses of this book has been to show this is possible. This 'opening of the doors of perception' can be drug-induced altering our normal perception chemically. Here the emphasis has been on how our perceptual range can be widened by less drastic means. This can be done by appreciating how much perceptual abilities actively construct the world in addition to being passive receptors of sensation and by not relying so exclusively on our *visual* perception.

James Gibson, in his book *The Senses as Perceptual Systems*, demonstrates convincingly how the senses are not just channels for receiving sensations but active systems in their own right and to some extent independent of sensation. "Sensation is not a requisite of perception and sensory impressions are not the 'raw data' of perception – that is, they are not all that is given in perception" (Gibson: 1996, 48). The distinction between the active and passive elements of perception are acknowledged in everyday non-technical language. Listening is active hearing, feeling is active touching, observing active seeing. Senses are not just passively responding, they are actively detecting and so seeking out what shall be experienced and just as importantly, filtering out the masses of input *not* to respond to.

Understanding Time

Change can originate on the perceptual side. An unfamiliar sensation, whether it be auditory, visual or tactile, can cause us to modify our view, prising us loose from our habitual thought patterns. In *The Doors of Perception* Aldous Huxley claimed that his experiences of the impression of colours under the influence of mescaline gave him insights into different concepts:

"To be shaken out of the ruts of ordinary perception, to be shown for a few timeless hours the outer and inner worlds, not as they appear to an animal obsessed with survival or to a human being obsessed by words and notions, but as they are apprehended, directly and unconditionally....this is an experience of inestimable value to everyone." (Huxley: 1959, 60).

Conceptual and perceptual powers match and supplement each other. With a more 'open' mind we are more likely to be aware of, even seek out and be responsive to other sources of perception, and find external confirmations which match the truths of less limited conceptions. Previous unintelligible perceptions which were ignored or filtered out become meaningful.

Out of the different perspectives of time described earlier it is now possible to arrive at an understanding of the nature of time which unifies the apparently different 'types' of time. The following paragraphs bring together threads from the preceding chapters, weaving them into a coherent pattern to support a clearer view of the 'nature' of time, in which there is no longer any need for the spurious distinction between an objective 'real' time and a subjective 'mind' time. Time as a living heterogeneous flow of experiences and time as a passive homogenous collection of measurable units of seconds, hours, minutes and days are no longer irreconcilable, but different aspects of the same phenomenon – time.

What is Time?

In the chapter on philosophy it was described how Kant maintained human cognition was limited by our spatio-temporal intuitions, so that any experience which was not filtered through these conditions could not be known. Kant held that behind that world as perceived lay the causes of the perception which Kant christened the 'noumena', unknowable causes of our sensation behind the phenomena, which we 'intuit' (sense) in space and time. They can never be known because we can never step outside our own perception. However, Kant did concede there was something he described as the 'synthetic unity of apperception'. This is somewhat akin to the modern neurological 'protoself', which does function before spatio-temporal intuition, but cannot give rise to any knowledge, as it has no content on which to work. But some philosophers, notably phenomenologists, hold that we do know immediately, before we cognise, this precognitive knowledge being unconditioned by our normal spatio-temporal perception. After that the subsequent chapter looked at how the world is experienced non-philosophically.

In everyday experience it is self-evident that every moment is unique and unrepeatable. Yet in the processes by which those experiences are cognised their uniqueness is weakened. Without this there could be no conceptual understanding which can only work by placing the particular under the general, classifying the myriad heterogeneity of experience under categories. It is only the widespread influence of Western thought based on Greek philosophy that has smothered this distinction, by elevating the general concept as somehow more real than the individual particular event.

The uniqueness of each moment is because human consciousness is not just a passive observer of external events.

Understanding Time

Our visual perception cannot be replaced by a rewindable CCTV camera film any more than our hearing can be replaced by a recording device. This is obvious intuitively but can become complicated when it is being conceptualised. Nevertheless, it has to be conceded that our awareness of events depends on our intentions reacting to and interpreting stimuli, by no means confined to the visual. Similarly our means of expression relate to many other faculties besides the conceptual: "Systematic reasoning is something we could not, as a species of individual, possibly do without. But neither, if we are to remain sane, can we possibly do without direct perception... of the inner and outer worlds into which we have been born. This given reality is an infinite which passes all understanding and yet admits of being directly and in some sort totally apprehended – to be enlightened is to be aware, always, of total reality and yet to remain in a condition to survive as an animal, to think and feel as a human being, to resort whenever expedient to systematic reasoning." (Huxley: 1959, 63-4)

How the immediate pre-cognitive data of consciousness although not verbally expressible can be expressed in free action, was then examined in the chapter on freedom. This concentrated on Henri Bergson's work. For him, in free acts our self can use an internal sense of time, where prior to spatial externalisation, past, present and future are not necessarily discrete. In this preconceptual flux, Bergson's 'la durée', prior to or independent of a spatial time, actions can be freed from the deterministic chain.

In the next chapter how freedom is realised in creative acts was examined, particularly in art. In the pre-conceptual world it is not only time that is not separated into discrete units. The division between the external and internal worlds, between the knower and known is equally fluid. The artist comes so close to the work that distinguishing the creator from the created becomes more and more difficult. The Kleistian notion of grace in which the act and the actor become one, where there is no spatio-temporal separation of their essential unity, becomes paramount. The integral unity of a work of art was also explored, in particular, Zuckerkandl's notion of a temporal gestalt. His thesis that the individual part does not acquire its meaning only in itself but from a whole can be extended beyond his application solely to music. Temporal unity precedes spatial unity even in awareness of the plastic arts which are primarily spatial. This is because any awareness must belong to a self-conscious living organism in time. Space is always contingent, a partial and fixed externalisation of time, though necessary for any external awareness, even awareness of ourselves as objects. The world is mirror to the mind and so the viewer never sees the unreflected self behind, but only its projection.

Time and consciousness are inseparable, but nevertheless distinguishable. There is a whole new scientific 'nature of consciousness' debate, whereas in earlier decades it had been seen as outside science's terms of reference. There is an 'Association for the Scientific Study of Consciousness' as well as a *Journal of Consciousness Studies*. In the hands of some this means that science can simply reduce consciousness to another area in which science alone has the authority:

"One of the striking, even amusing, spectacles to be enjoyed at the many workshops and conferences on consciousness these days is the breath-taking overconfidence with which laypeople hold forth about the nature of consciousness, their own in particular, but everybody's by extrapolation. Everybody's an expert on consciousness, it seems, and it doesn't take any knowledge of experimental findings to secure the home truths these people enunciate with such conviction. One of my goals over the years has been to shatter that complacency, and secure the scientific study of consciousness on a proper footing." (Dennett: 2002) What a risible conceit!

But, as shown in the chapter on the New Physics, not all scientists want to hijack consciousness as they have done with time. With the dualistic separation of the observer and the observed no longer tenable and rejected by the new physics, forms of monism are becoming acceptable. This does not have to be the materialist monism as favoured by Dennett and Dawkins; the alternative of idealistic monism has its proponents, most notably and controversially, Goswami. Idealistic monism does allow for a genuine dialogue between science and other disciplines whereas materialist monism tends to regard the only truths to be scientific ones.

For science on a superficial level, being in what could be called a 'preconceptual' stage of knowing which could escape the limitations of our physical time perception presents a problem. This is because the primacy of *ex-post facto* perception, which is science's 'stock in trade' loses its exclusive position as the source of all knowledge. This is worrying to those scientists, and there are many, who believe that knowledge must solely be based on an interpretation of objective reality as experienced through conscious perception.

But, for 'instrumentalists' this presents no problem. Science is not concerned with any absolute truths; it is concerned with what works in a particular sphere or range. They are happy with what Nancy Cartwright describes as a 'dappled world' where the theories are relative to their functionality and only apply within a circumscribed sector. Normal scientific methods based on the experimental method are totally valid within their areas of application. In its practical pursuit of knowledge of the objective world generally, but not always, science can legitimately ignore the effects of consciousness as insignificant. They can safely be disregarded and a *ceteris paribus* principle applied, hermetically sealing off one particular world from the others. Scientific 'realists' on the other hand reject this approach. Even where it is accepted orthodoxy, as in the application of quantum physics to the minute sub-atomic world and of relativity to the incredibly large, they balk and want a unified theory which combines two seemingly contradictory interpretations of nature.

In sum then, the question 'What is time' is a 'category mistake'. Time is not a thing we perceive and then form a concept of which one asks the question 'What is it?' Plato's description of time in the *Timaeus* as an image is closer. "The moving image of eternity" expresses the nature of time, when viewed not as a definition but more as a poetic paradigm in which different senses of time can be placed, as our consciousness widens or deepens. How the image is felt will vary as we can never completely step outside our perceptual apparatus, habits and background. But the more we disentangle ourselves from them, the more we are aware of a less limited time.

For different consciousnesses or states of consciousness there are different times. These may be argued to be all different partial awarenesses of some absolute time, but this is rather like clinging on to the search for the one unifying complete theory of everything . For us, therefore, time a not an object of perception, it is a condition, and therefore a limitation, but it is a condition which can be altered.

Our cross-section of the present is Ouspensky's 'slit', which contains a moving image of eternity. However much we widen the slit we will still be in a world of past, present and future. 'Eternity' must be when there is no 'slit'. This is a time unbounded by perceptual and mental constraints and therefore in which the feeling of 'now' loses its meaning as it is no longer limited to a perceived present. This 'pure' time is part of the noumenal and therefore the paradox of whether it is outside in the world or inside our minds dissolves 'From the point of view of eternity time in no way differs from...extensions of space – *length*, *breadth* and *height*. This means just as space contains things we do not see; or to put it differently, more things exist than those we see, so in time 'events exist before our consciousness comes into contact with them, and they still exist after our consciousness has withdrawn from them'. (Ouspensky: 1981, 33).

The question is how do they 'exist'? Common sense tells us the external reality must exist even when no one is observing it. Nevertheless, time is not a property of objects; it is always relative to a consciousness which is presupposed. Without consciousness there can be no time. Bishop Berkeley's solution was God, keeping an idealist world in being when no other consciousness was maintaining it. This is mocked in the following limerick:

There was a young man who said "God
Must think it exceedingly odd,
If He finds that the tree,
Continues to be
When there's no one about in the quad."

" Dear Sir, your astonishment's odd;
I'm always about in the quad.
And that's why the tree
Continues to be,
Since observed by yours faithfully,
– God."

But there are other supports for the assertion that external reality does not exist when not observed; it does not have to rest on the assumption of an anthropomorphic God. Since the advent of quantum theory the firm belief of science in an external reality has wavered and therefore its incompatibility with and intolerance of non-scientific approaches for some scientists is lessened. Indeed, for Fritjof Capra the findings of modern physics are close to those of Chinese and Indian mystics, who arrived at the same conclusions by a totally different route:

"Science, it is usually believed, helps us to build a picture of objective reality – the world 'out there. With the advent of quantum theory, that very reality appears to have crumbled, to be replaced by something so revolutionary and bizarre that its consequences have not yet been properly faced….one can either accept the multiple reality of parallel worlds, or deny that the real world exists at all" (Davies: 1980, 12)

One common interpretation of reality by modern physics already encountered is Neil Bohr's 'Copenhagen Interpretation' of quantum theory. By this when no one is watching, a virtual system changes deterministically according to a wave equation. But only when a conscious being apprehends it does the wavefunction of the system "collapse" to an observed 'real' state. By this it is consciousness that *causes* a quantum 'collapse', which calls a world into being. Indeed according to Wigner's ideas discussed in previous chapters without intelligent conscious life there is no external world, which is a quite uncompromising idealist position.

The world is forever new, actualised by a consciousness's time sense. This indeed is what things feel like before our reason tells us the contrary. Time is the foundation of both internal and external reality. All human consciousnesses agree broadly on their version which they confuse with the 'true' version, as we can be directly conscious of no other. This external world only exists virtually in potentia until it is made immanent by consciousness. The quantum collapse of the immanent world made *manifest* depends on the nature of *the* consciousness which intuits it. And the nature of this consciousness varies from species to species, from culture to culture and crucially from individual to individual.

At this level the question about the nature and meaning of time becomes inextricably linked with that of consciousness. This was inevitable as there can be no time without consciousness and no consciousness without time. Time is not an "extra" dimension to space. It is the ground of our individual being, creating the fundamental lacuna between sentient and the sensed. By contrast, space creates the possibility of externalising our internal awareness into an external reflection of it through which objective knowledge is acquired. Into this objective world we conceive an 'objective' image of time.

But this conceptualized objective time is a time without self-consciousness. It is in fact an absence of time, no time at all, a time reduced to space – a dead eternity of no time in a world without motion or flow. In contrast there *is* a living eternity, the 'eternal now' of an unbounded consciousness. Internal time is Plato's ever-changing moving image of eternity; spatial time is just a frozen snapshot.

Hence the root fundamental paradox of time. A conceived time is always a time from which consciousness has been extracted to make time homogenous. But a self-conscious time which must precede any conceptual framework can only be communicated conceptually by distorting it. So there is a danger of returning to where the discussion started with St Augustine's dilemma. "If no one asks me I know; if someone asks me to give an account, I don't know".

But this is only if it is accepted that communicable knowledge is limited to the conceptual knowledge. That there are ways of communicating non-conceptual knowledge has been shown earlier in this book. These cannot be objective because that requires extension of time in the homogeneous medium of space; but neither can they be totally subjective, in the way that no one can directly experience someone else's pain or wonder. They have to be inter-subjective, where the truth or meaning is conveyed by arousing a response based on shared forms of consciousness, rather than an ability to cognise by conceptual understanding.

Experiences or events which are known with any spatial fixed point of reference are non-local. The concept of non-locality has been extended beyond its use in quantum physics. Gilles Deleuze, as was seen earlier in the chapter on Art, applies it to cinema theory. F. David Peat, a close colleague of David Bohm, whose theories were looked at in the chapter on the New Physics, considers non-locality provides a complementary approach which is relevant to all knowledge. "Non-locality could be considered as a complementary description to that of locality, as part of a general nexus of new ideas." (Peat: 1991, 12) In particular, he applies it to the working of the brain, art forms, especially music, and communication theory:

"By seriously considering the ideas of non-locality, not only in quantum theory but in a much wider context, it may be possible to develop a conception of nature that integrates more deeply with our own perception, thought and experience" (Peat: 1991, 7)

Non-locality is felt or experienced in time alone. Some, like Dostoevsky, who think that time is essentially unreal see it as a product of consciousness and therefore a subjective internal time. Others, such as Gilles Deleuze, believe it to be more 'real' than spatial time which distorts and limits it and call it 'pure' time. But whether it is real or ideal is in a sense irrelevant.

Whatever its ontological status, experience of non-locality is only possible when consciousness is 'delocalised'. This can be achieved in meditation by 'withdrawing from the world' and in acts of creation, artistic or otherwise. Art communicates these experiences non-verbally through visual or auditory images which provoke a non-local response. The mystical experience can also be communicated if words are used in a non-local way, as images.

Silence, which could never be used to communicate conceptual knowledge can be a powerful medium. In one-to-one situations between lovers or an inspired teacher and pupil a soundless act or just eye contact can communicate a shared thought or emotion. On a general inter-subjective level the use of the rest in music is an example.

To conclude, either words must be used differently (perhaps more poetically) to express a known and experienced but inconceivable time, or non-verbal means of expression must be invoked. Perhaps only non-verbal knowledge can approach the essence of time as words, even when used as poetic images, must inevitably create a blur between the world as known and the world as it is subsequently described. The techniques and symbols of music and mathematics seem to be able get closer to the essence of time and consciousness than the prosaic medium of prose.

Chapter Seven

Metatime

"If the doors of perception were cleansed every thing would appear to man as it is, infinite. For man has closed himself up, till he sees all things thro' narrow chinks of his cavern. . ." *The Marriage of Heaven and Hell* William Blake

Metatime is used here to signify a time beyond our normal awareness when, in science, art, everyday life and philosophic thought and actions, the limitations of 'our' time and therefore our space within that time, are transcended. There is another sense in which the term 'metatime' has been used to describe a 'second' time by which the flow of time is measured. But this leads to an infinite regress as this second time must have a third time to measure *its* rate of flow and so on ad infinitum. But we are not dealing here with an external time which can be measured. This is about an internal time unrelated to any notions of infinite regress.

This transcendence involves a development of 'our being in the world', either mental, physical or emotional, by which our mind, feelings or body, often an indivisible fusion of all three, have a new 'higher' kind of relation with the world.

Heightened mental awareness can come naturally. It comes about in moments of calm or extreme stress. In crisis, time appears as slowing down, even stopping. It is therefore in a sense internal, giving the mind the 'time' to react without panic. When all is peaceful and fulfilled the feeling of the passage of time sometimes disappears; everything is so perfect that it is as if time were suspended.

As for heightened emotions, falling in love opens up in people sides of their nature they were not aware of as can a highly significant impression or event. Again the passage of time can seem slower as we have a reunion with a friend or watching at a sick child's bedside.

Physically, our bodies can often feel a kind of time-transcending ecstasy when we do something with a perfect grace, for example in a perfect physical performance, in sensual closeness or in the creation of a work of art which involves close physical work such as pottery, sculpture or architecture.

But these changes of awareness can be cultivated as well as being spontaneous. This can be done through different human faculties. Using the classification of the Bhagavad Gita there are three different paths of yoga: jnana, bhakti and karma, which correspond to the three kinds of emphasis of a human personality – contemplative, emotional and active respectively. Adult development can have different levels and directions depending on the individual and roughly correspond to these divisions, though of course this is only an ad hoc classification as everyone is partly contemplative, party emotional and partly active – it is just a question of which predominates. The Buddhist eightfold path (right view, right intention, right speech right action, right livelihood, right effort, right mindfulness and right concentration) recognises this. In fact in the way to enlightenment of the Buddhist 'Middle Way' each of these traits must be developed. Taken as a whole they encompass our emotional, intellectual and practical nature. Each of the three paths is described separately, always bearing in mind, however, that they are intimately linked.

Contemplation

In the contemplative personality the path has been described as "a movement of consciousness towards a higher level, as a result of the emergence and cultivation of powers which in most men or women remain latent. As found in the true contemplative it is an extreme form of the withdrawal of attention from the sensible world". (Happold: 1963, 69) Throughout the earlier chapters, there has been a theme of pre-conceptual knowledge. This has not been caught in the 'veil of (spatio temporal) perception' because it has not yet descended. There is, though, another way of escaping the limits of perception and moving into a different non-cognitive time. This is by shutting off our consciousness from the external world – the mystic way, transcending time and space by withdrawing into the internal. This is not pre-conceptual, more non-conceptual or extra-conceptual, but the state of mind is similar, approaching that of the inspired artist. It is "an extreme form of the withdrawal of attention from the sensible world and a total dedication of action and mind towards a particular interior object... essentially a creative activity, similar to the highest activity of the poet, painter and musician. All great creative artists, whether their medium be words or paint or sound, must be in a real sense illuminated, and in some is clearly seen that remaking of consciousness which is found in the religious contemplative" (Happold: 1963, 69)

Applied to time, those who have stepped back from the world begin to sense that external time is an illusion upon which all our ideas of a permanent external world rest. In an uncompromising fashion Buddhism considers that all concepts including the sense of self are Maya, an attempt by the mind to reduce the everchanging and indivisible flow of reality into discrete categories. This may be very useful, maybe essential, in practical life, but the error is to mistake this for reality and thus deny the contemplative's truth that all is one when the sensible world is transcended. It is not just Buddhism that emphasises this and not just 'religious' thought. Coomaraswamy, in his book *Time and Eternity*, documents how much similar insights permeate Islamic, Greek and Western thought, whether religious or secular, as well as Buddhist and Hindu thought, with their Vedantic precursors.

In the 4th century BC Aristotle wrote the 'Organon' setting out the principles of logic. In 1620 Francis Bacon published the *Novum Organum* which abandoned Aristotelian deductive logic in favour of the experimental method of modern science. In 1912 Ouspensky, a Russian mathematician, journalist and mystic published a work with the presumptuous title of *Tertium Organum*. 'Tertium Organum', he claimed, "*for us* is the third instrument of thought after Aristotle and Bacon. The first was ORGANON, the second NOVUM ORGANUM. But the third existed before the first." (Ouspensky: 1981, 221). As his language is more accessible to the present age this section will draw heavily on his work, though he too has the same difficulty in expression as his predecessors:

"Everything said about a new understanding of time relations is bound to be very obscure. This is so because our language is entirely unadapted to a *spatial expression of time concepts*. We have not got the necessary words for it, we lack the verbal forms. Strictly speaking, the expression of these relations, new for us, require some quite new, different language, perhaps *non verbal*. Until then, in our human language, we can speak of 'time' only by hints. The true essence of it is *inexpressible* for us. We must never forget this inexpressibility. *This is the sign of truth*, the sign of reality. That which can be expressed cannot be real. All systems speaking about the relation of the human soul to time...all these are symbols, striving to transmit relations which cannot be expressed *directly* owing to the poverty and weakness of our language. They should not be understood literally, just as one cannot understand literally artistic symbols and allegories." (Ouspensky: 1981, 98)

The contemplative's inner world is somewhat analogous to Kant's noumenal world, a world outside experience as conditioned in space and time:

"*Duality* is the condition of *our* perception of the phenomenal (three-dimensional) world; it is the *instrument* of our perception of phenomena...but when we come to the perception of the noumenal world... this duality stands in our way.. We, in order to understand...must renounce the *idol of duality* ..." (Ouspensky: 1981, 224)

The opposite of dualism is monism. Ouspensky's assumption that there is something beyond the phenomenal world evidently means that he is advocating a monistic idealism rather than monistic materialism.

Ouspensky summarises the difficulty of communicating the inner sense of oneness which the contemplative experiences: "but the application of monism to practical thinking comes up against the insurmountable obstacle of our language. Our language is incapable of expressing the unity of opposites." (Ouspensky: 1981, 224).

This attitude is shared by many profound thinkers but there is the usual obscurity caused by expressing in language what may lie beyond language. It is present in many well-known mystical and philosophical statements. For instance the *Thou art That* of Hinduism only means something to the initiated. Then there is the famous quote from Wittgenstein "That of which we cannot speak, thereof we must remain silent" and a similar denial in Lao Tzu "Those who speak do not know, those who know do not speak." That of which we cannot speak, thereof we must remain silent.

Ouspensky, however, believed that art may be the basis for a new language by which these insights can be communicated. This connects closely to the previous chapter in which it was stressed that the use of words in the poetical or analogical sense, which is it itself artistic, may be one way to communicate the contemplative's insights.

Poetic words like other art creations are, nevertheless, only pointers to something which has been experienced. If you haven't had direct experience the artistic symbols cannot be recognised. One of the most celebrated expressions of this requirement is Plato's allegorical Dialogue of the cave in the Seventh book of 'The Republic'. If the unchained cave dwellers returned from the surface to bring the good news to the chained dwellers that the fire-illuminated shadow puppet show, which is all they know, is false, they would be met with a complete lack of understanding. In fact, Plato goes further, writing that a more likely reaction would be that the chained 'prisoners' would think that the unchained returnees had lost their abilities to appreciate the shadow puppet due to their exposure to the light and that they would want to put to death anyone who tried to unchain them.

Another way of attempting to express the transcendent in words is by using a kind of negative logic. Instead of trying to describe what may be intrinsically indescribable in our current language, some have attempted to prove that there are parts or indeed the whole of reality that can *never* be described in words. Kant's 'Critique of Pure Reason' shows the limits of our cognitive abilities. He demonstrates the tangles human reasoning power gets in when it exceeds its limits. Our phenomenal world is always mediated by our human species' particular set of spatio-temporal 'spectacles'. The unknown and unknowable causes of phenomena will always be beyond our knowledge because we cannot step outside our human form of spatio-temporal perception to experience them. Personally I have always felt uncomfortable with noumena and feel that Kant's system is much clearer without them. The mystics of many religions tend to describe the nature of truth in similar negative fashion but without invoking the concept of noumena. "The Tao which can be spoken is not the eternal Tao" (Lao-Tzu) is a lot simpler.

Stepping back from the contemplatives' experience, where does that leave the knowledge that natural science has revealed about the external world, which is applied every day in a much more practical way to affect our lives. The answer is, this is an entirely different kind of knowledge but that the two are not mutually exclusive and indeed may be an instance of how extremes coalesce. From Pythagoras to Newton to the present day, there have always been scientists who have combined a deep contemplative's insight with a genius for unravelling the material pattern of nature. Active and contemplative modes of being the world can both be present in one person. In fact they can mutually nurture each other. One develops an understanding of human nature which helps us attain happiness and fulfilment; the other helps us to use the world outside more generally to better our material needs. In the end they meet. This is the theme of Fritjof Capra's *The Way of Physics*, in which he quotes twentieth century scientists such as the quantum physicists Niels Bohr and Robert Oppenheimer.

" For a parallel to the lesson of atomic theory [we must turn] to those kinds of epistemological problems which already thinkers like Buddha and Lao Tzu have been confronted when trying to harmonise our position as spectators and actors in the great drama of existence" Niels Bohr 'Atomic Physics and Human Knowledge' quoted in Capra (1976: 16). Julian Oppenheimer similarly states that an atomic physics is an example of "a refinement of old wisdom".

This echoes Ouspensky's claim quoted earlier about the antiquity of his *Tertium Organum*.

Emotion

This is transcendence through Love in all its many forms – parental, sexual, universal and brotherly or sisterly. The classic Christian text on this path has to be *The Cloud of Unknowing* which is just a one-pointed message about how it is only through Love that we can attain union with God. Thought is simply a hindrance to be discarded on our journey upwards.

If the emotional path is followed transcendence is not thought but felt. The unity of the world is experienced as a perfect all-embracing love Juan Mascaro's *Introduction to the Upanishads* compares the message of love, as expressed in them over 2500 years ago, with examples from other religion's doctrines, and with poets and spiritual thinkers. This is now quoted at length:

"In this way.... the doctrine of the *Upanishads* explains and complements the Gospels, 'Thou shalt love thy neighbour as thyself'. Why? Because our Atman, our higher self dwells in us and in our neighbour: if we love our neighbour, we love God who is in us all and in whom we all are; and if we hurt our neighbour, in thought or in words or in deeds, we hurt ourselves, we hurt our soul: this is the law of spiritual gravitation.

Love is undefinable, but we know that love is joy: not indeed a transient pleasure, but an eternal joy of the soul......

A song of love is heard as a background to all great prayers. Chaitanya, the great Indian mystic. (A.D. 1500), pours out his heart in these words:

> 'I pray not for wealth, I pray not for honours, I pray not for pleasures, or even the joys of poetry. I only pray that during all my life I may have love: that I may have pure love to love Thee'

'Never fail, whatever may befall you, be it good or evil, to keep your heart quiet and calm in the tenderness of love'. With this one sentence St John of the Cross explains the doctrine of the *Bhagavad Gita*. When in the *Gita* we read again and again that a man must be the same in heat or in cold, in pleasure or in pain, in victory or defeat, the meaning is, of course, that whatever may be the events of our outer or inner life we must ever have the peace of love: in fact that our life should perpetually breathe the air of love, since love is the living breath of the soul. And far from an evenness of love making us insensitive, it is that love which leads to that sublime state described in the *Bhagavad Gita*

> 'And he is the greatest Yogi whose vision is ever one: where the pleasure and pain of others is his own pleasure or pain.' " (Mascaro: 1965)

Expressions of this love can be found in the mystics of the world's great religions. A common theme is to allegorise the unity of the Soul with God, or in less religious terms the one with the all as the love between man and woman, bride and groom. This occurs for example in Jakob Boehme, Saint Teresa and The Song of Solomon. Sufi poetry, in particular, is permeated with sensual metaphors. The renowned 13th century Islamic mystic Jalal ad-Din Muhammad Balkhi, better known in the West as Rumi, specifically describes love between human beings as a way to divine love

'Never will a Lover's chest
feel any sorrow.
Never will a Lover's robe
be touched by mortals.
Never will a Lover's body
be found buried in the earth.
To Love is to reach God.'

Rumi (translated by Shiva)

As all is undivided this feeling of boundless love extends to all living things. So we have Francis of Assisi preaching to the birds and the Buddhist precept to avoid harm to all sentient beings

It also embraces nature. In this heightened emotion, just as in heightened contemplation the duality of world and mind dissolves. From the early Vedas through to Western Romantic poetry, nature has been felt as a spiritual presence. As the duality lessens the border between the consciousness and its object becomes hazy.

Homme, libre penseur! te crois-tu seul pensant
Dans ce monde où la vie éclate en toute chose?
Des forces que tu tiens ta liberté dispose
Mais de tous tes conseils l'univers est absent.

Respecte dans la bête une esprit agissant:
Chaque fleur est une âme à la Nature eclose;
Un mystère d'amour dans le métal repose;
"Tout est sensible!" Et tout sur ton être est puissant.
Crains, dans le mur aveugle, un regard qui t'epie:

Metatime

A la matière même un verbe est attaché...
Ne la fais pas servir à quelque usage impie!

Souvent dans l'être obscur habite un Dieu caché;
Et comme un oeil naissant couvert pas ses paupières,
Un pur esprit s'accroît sous l'écorce des pierrres!

Vers Dorés Gérard de Nerval 1854

A free translation is as follows:

Free Man! Believe you the only being with thought
In this world where life bursts out in all things?
The power that you hold your liberty brings
But in all your plans the universe counts for nought.

Respect in the beast a spirit which is stirring,
Every flower is a soul to nature exposed,
A mysterious love is in metal reposed,
"All is sentient" and has power over your being.

Fear, in the blind wall, a gaze which descries one
Even to matter a verb appends
Do not enslave it for impious ends

A pure spirit is growing under the crust of a stone
Like an eye being born still concealed by its lid
In obscure beings often a God is thus hid!

The later Wordsworth writes in a similar vein:

"One impulse from a vernal wood
May teach you more of man,
Of moral evil and of good,
Than all the sages can.
Sweet is the lore which nature brings;

Understanding Time

Our meddling intellect
Mis-shapes the beautous form of things:–
We murder to dissect.
Enough of Science and of Art;
Close up those barren leaves;
Come forth and bring with you a heart
That watches and receives"

'The Tables Turned' William Wordsworth

Poetry is sometimes better at expressing deep emotion than prose but I shall end this section with a prose quotation from one of the greatest of modern poets:

"And when Love speaks the voices of all the gods, Makes heaven drowsy with the harmony", without it our existence is just like a 'tinkling cymbal, 'love conquers all', "love all alike, no season knows or clime, Nor hours, days, months, which are the rags of time" ...literature is full of references to the transcendent power of love. Earthly sensual love is often taken as a metaphor for a more rarefied love. It is the yearning for or achievement of a blissful union. Requited or unrequited, it releases an overflowing intensity which only an eternal sense of time can match.

"Around those who love is sheer security. No one casts suspicion on them anymore, and they themselves are not in a position to betray themselves. In them the secret has grown inviolate, they cry it out whole like nightingales, it is undivided. They make lament for one alone, but the whole of nature unites with them: it is the lament for one who is eternal. They hurl themselves after him they have lost, but even with their first steps they overtake him, and before them is only

God. Theirs is the legend of Byblie, who pursued Caunus as far as Lycia. The urge of her heart drove her through many lands upon his track, and at last she came to the end of her strength; but so strong was the mobility of her nature that, sinking to earth, she reappeared beyond her death as a spring, hurrying on, as a hurrying spring.

What else happened to the Portuguese nun, save that inwardly she became a spring? Or to you Héloise? To you all, lovers, whose laments come down to us.." (Rilke: 1949, 198)

Action

As in contemplation the separation of the self from the world is transcended, so in inspired action the separation of the performer and the action performed is transcended. In English for both these states the word 'grace' is used to describe the feeling of deliverance or (atonement) at-one-ment from the tension of waking consciousness in which the self and the objective world are held apart. The physical expression of this grace has many forms. It can happen in nearly any action with which a person can become totally identified ranging from the profound and fateful, skilled pursuits, physical or emotional achievements down to the most mundane acts.

Sport is a very good example. 'Grace' is often used to describe a superlative performance in ice-skating, gymnastics or for a dramatic goal or save in soccer. It applies equally to the sportsperson's inner feeling at the time as well as his or her outward performance. For a moment of Grace body and mind become one harmony for the athlete, ballet dancer or footballer.

Often this unity extends to the sporting material with the player 'merging' with the apparatus: the tennis player with the racquet or the discus with the thrower.

Another way of action is artistic creation. The sense of time transcendence experienced by the artist as he or she creates, or when the work is appreciated, has been described earlier in Chapter 4, as a kind of virtual world in which art is an emotional understanding rather than a mental understanding. In this virtual world everything is not what it is thought or observed as, but what it is felt as. Analysis of a painter's tools or works with a spectroscope or chemical agents will reveal nothing of their meaning; similarly the dictionary definitions of words in a poem will accomplish as little as its deconstruction. A different kind of consciousness is invoked in which the spatio-temporal divide between the artist and the work becomes blurred and virtually vanishes, giving that artistic grace, which in the chapter on creation was exemplified by Kleist's description of puppetry.

In sexual relations, the physical and the emotional are conjoined. The feeling of transcendence in sexual union does not necessarily depend on any creative or physical perfection. The power of sex is recognised in many great systems of thought whether religious, philosophical or scientific. Its transcendent powers are sometimes seen as a threat. But, just as often it is treated as another way of transcendence, especially in some Eastern traditions such as Tantra Yoga. It is no accident that men of action and great artists often have prodigious sexual drives.

Then there is action on the grand scale often linked to a feeling of destiny in which the future now the present seems to have been preordained. When time stands still we can exclaim "Stay Stay moment thour't so fair" '(Goethe, *Faust* Part II), as Faust does not through the fulfilment of his self-indulgent wishes, but through his land drainage schemes. The same sentiment is present in the Bodhisattva ideal of Mahayana Buddhism, "Enlightened Beings who renounce Nirvana's bliss in order to remain in the universe and aid the liberation of their fellow beings" (Blofeld: 1970, 65)

At the other end of the spectrum in Zen Buddhism (both traditional and as interpreted by the West) there is an emphasis on how even the most mundane or trivial act can be transformed. As examples, on the traditional side there is the art of flower arranging; in the West it is present in Robert M. Pirsig's *Zen and the Art of Motorcycle Maintenance.*

In the end even the smallest action can trigger a time-transcending experience if it is inspired with the whole self. These can be acts heavy with symbolism such as eating a communion wafer, but can be extended to any action from gardening to listening to music or changing a baby's soiled clothes. The physical act itself, just like the artists' tools, can be accompanied by conscious intention and meaning. Jakob Boehme was an ordinary shoemaker, certainly no intellectual heavyweight, who experienced and recorded his illuminations with such power that he is still read with respect today. The external circumstances which acted as catalysts for him were quite ordinary. The first was the reflection of sunlight from a pewter dish and the second sitting down in a field during a walk. In his third illumination, in which he states he learnt more than if he had been many years at university, its duration was just one paltry quarter of an hour.

Being in Metatime

Getting beyond our everyday working knowledge and understanding of time is a development reached by different character types in different ways. As mentioned earlier they are all linked and analysing them as intellectual, emotional or active is somewhat artificial. Contemplation usually results in a 'higher' emotional state which can only with difficulty be separated from the mental experience. Similarly, heightened emotional states often precede or accompany inspired action. It is therefore unsurprising that they all share the same essential characteristics.

One of these is that the change is not transmittable by explanation from a developed individual to a less developed individual. It has to be experienced so the 'lower' can never know the 'higher', though of course the 'higher' still knows the 'lower' having been through that stage earlier. Just as a child cannot learn what mature adult love is like, the insights of a person of deep knowledge or exceptional emotional depth cannot be copied by following a course of instruction. The nearest approach is when Zen masters attempt instantaneous enlightenment by shock tactics, but even they concede that this only works if the student has already almost attained the level.

Another is a feeling of oneness, lack of separation between the self and the world. As the spiritual side of life predominates there are glimpses of how the human spirit can eventually merge into union with the transcendent or the divine. Specific differences between those on a similar path become unimportant. The separation of our own existence from everything else, the subjective from the objective, the self from the non-self, and mind from matter, become irrelevant or meaningless. Everything becomes one, yet at the same time a vestige of individuality remains.

As this occurs our time sense becomes less focused. The 'slow time' or 'Dreamtime' of 'unWesternized' cultures is recaptured which has been lost in a culture obsessed by speed and efficiency.

"It is a culture ignorant of the past and viciously refusing to plan for the future, not respecting the young, nor respecting the old. Its exports are adolescent: fast cars, fast food, fast talk.... modern westernized cultures could never have produced the *Kama Sutra*, would never pause to consider the point of orgasm maintained for hours" (1999: Griffiths, 41)

This links to another characteristic. Though the time spent in these rarefied states may be fleeting, the intense quality of the experience is felt as a longer quantitative experience, There is a natural tendency to exaggerate the 'clocktime' duration, William Blake's "feeling eternity in an hour".

The flow of normal self-conscious time is often given the metaphor of a river; metatime is often analogised as an ocean into which rivers flow. In the ocean there are no longer individual 'streams of consciousness', yet even here some sense of individuality is maintained. The rest of this chapter will rely quite heavily on poetry as I try to portray this flow.

Rumi the Sufi poet expresses this succinctly." Many understand that the drop exists within the ocean but few can comprehend how the ocean can exist within the drop ..".

Understanding Time

Like Rumi, Walt Whitman also uses the medium of water:

"As I wend to the shores I know not,
As I list to the dirge, the voices of men and women wreck'd,
As I inhale the impalpable breezes that set in upon me,
As the ocean so mysterious rolls toward me closer and closer,
I too but signify at the utmost a little wash'd-up drift,
A few sands and dead leaves to gather,
Gather, and merge myself as part of the sands and drift.

O baffled, balk'd, bent to the very earth,
Oppress'd with myself that I have dared to open my mouth,
Aware now that amid all that blab whose echoes recoil upon
me I have not once had the least idea who or what I am,
But that before all my arrogant poems the real Me stands yet
untouch'd, untold, altogether unreach'd,
Withdrawn far, mocking me with mock-congratulatory signs
and bows,
With peals of distant ironical laughter at every word I have
written,
Pointing in silence to these songs, and then to the sand
beneath.

I perceive I have not really understood any thing, not a single
object, and that no man ever can,
Nature here in sight of the sea taking advantage of me to dart
upon me and sting me,
Because I have dared to open my mouth to sing at all.

Metatime

You oceans both, I close with you,
We murmur alike reproachfully rolling sands and drift, knowing
not why,
These little shreds indeed standing for you and me and all.

You friable shore with trails of debris,
You fish-shaped island, I take what is underfoot,
What is yours is mine my father.

Kiss me my father,
Touch me with your lips as I touch those I love,
Breathe to me while I hold you close the secret of the
murmuring
I envy.

Ebb, ocean of life, (the flow will return,)
Cease not your moaning you fierce old mother,
Endlessly cry for your castaways, but fear not, deny not me,
Rustle not up so hoarse and angry against my feet as I touch
you or gather from you.

I mean tenderly by you and all,
I gather for myself and for this phantom looking down where
we lead, and following me and mine.
Me and mine, loose windrows, little corpses,
Froth, snowy white, and bubbles,
(See, from my dead lips the ooze exuding at last,
See, the prismatic colors glistening and rolling,)
Tufts of straw, sands, fragments,
Buoy'd hither from many moods, one contradicting another,
From the storm, the long calm, the darkness, the swell,

Understanding Time

Musing, pondering, a breath, a briny tear, a dab of liquid or soil,
Up just as much out of fathomless workings fermented and thrown,
A limp blossom or two, torn, just as much over waves floating, drifted at random,
Just as much for us that sobbing dirge of Nature,
Just as much whence we come that blare of the cloud-trumpets,
We, capricious, brought hither we know not whence, spread out before you,
You up there walking or sitting,
Whoever you are, we too lie in drifts at your feet.
                    As I ebb'd with the Ocean of Life." Walt Whitman

As long as we remain self-conscious a heightened awareness of time, 'metatime', is experienced as Bergson's "la durée". Bachelard in his book *L'Intuition de l'Instant* vigorously opposed Bergsonian flow. He held that such a flow can only have an indirect force. The only real time is the instant which is complete in itself, even if it may contain resonances like echoes, of past instants which no longer exist (1935: Bachelard, Section V). This is close to the Eternal Now of Indian philosophy in which separation into past, present and future is not applicable. If there is only an instant there can be no flow. If the only 'real' unit is the instant this can be interpreted as an acceptance that time is not 'real' (whatever that means). Julian Barbour in the Epilogue to *The End of Time*. Writes: "Each experienced instant is a separate creation (birth), the ever inaugural act of existence, brought to life by the gathering of all times" (Barbour: 1999, 334.) Whether the reality of the instant is defined as 'a cessation of time', an 'Eternal Now' or a 'separate creation' is not that important and seems to just be quibbling with words. But how long are instants?

In the Bergsonian flow, past, present and future can mingle and interpenetrate as our inner self is working in an immediate intuition of time before it is ordered sequentially. This durée is loosely analogous to the "specious" present characterised by William James, defined by Le Poidevin as "the interval of time such that events occurring within that interval are experienced as present" (Le Poidevin: 2000). It is, however, not a precise quantity. Like Ouspensky's 'slit' it varies from species to species and in the case of man it can vary intentionally or physiologically from person to person. Bergson's portrayal of a free act at crucial moments of decision could occur in this 'specious present'. Memory (conscious or unconscious), the present (the lag between the start of neural events leading to consciousness and the moment one experiences the contents of consciousness as a mental 'event') and anticipation of the future all merge.

Different from the "specious" present is the absolute present. This cannot have any duration. As St. Augustine maintained it must be durationless, because if it has duration there is an earlier and a later. The instant, as such, is beyond self-consciousness as there can be no minimal duration for the self to separate itself from the sensation. The instant is at the absolute extreme of Metatime. At its limit *metatime is not the time of self-consciousness, it is the time of consciousness stripped of self.* It follows that any such experience of absolute metatime cannot be of permanent duration and can only be known when self-consciousness returns – it is an ecstasy.

Understanding Time

Paradoxically there appears to be an affinity with the absolutes of mathematics and logic and the sensation of metatime. Both depend for their certainty on the removal of time as a variable. But this is the end of the similarity. In the case of mathematics when it is applied to the external world what is true is true if identical conditions are met; the direction and flow of time can be ignored. In the case of metatime what is true is true because the direction and flow of time are transcended. To describe this is almost as difficult as verbalising a mathematical formula, but the following poem does give a 'feel' of what it is:

Time was away and somewhere else,
There were two glasses and two chairs
And two people with one pulse
(Somebody stopped the moving stairs):
Time was away and somewhere else.

And they were neither up nor down;
The stream's music did not stop
Flowing through heather, limpid brown,
Although they sat in a coffee shop
And they were neither up nor down.

The bell was silent in the air
Holding its inverted poise –
Between the clang and clang a flower,
A brazen calyx of no noise;
The bell was silent in the air.

The camels crossed the miles of sand
That stretched around the cups and plates;
The desert was their own, they planned
To portion out the stars and dates:
The camels crossed the miles of sand.

Metatime

Time was away and somewhere else.
The waiter did not come, the clock
Forgot them and the radio waltz
Came out like water from a rock:
Time was away and somewhere else.

Her fingers flicked away the ash
That bloomed again in the tropic trees:
Not caring if the markets crash
When they had forests such as these,
Her fingers flicked away the ash.

God or whatever means the Good
Be praised that time can stop like this,
That what the heart has understood
Can verify in the body's peace
God or whatever means the Good.

Time was away and she was here
And life no longer what it was,
The bell was silent in the air
And all the room one glow because
Time was away and she was here.
                'The Meeting Point' Louis MacNeice April 1939

With time suspended, consciousness is at one with the infinite:

"External objects present us only with appearances. Concerning them, therefore, we may be said to possess opinion rather than knowledge. The distinctions in the actual world of appearance are of import only to ordinary and practical men...The subject surely cannot *know* an object different from itself.... Consciousness, therefore, is the sole basis of certainty...

"Knowledge has three degrees – Opinion, Science, Illumination. The means or instrument of the first is sense; of the second dialectic; of the third intuition. To the last I subordinate reason. It is absolute knowledge founded on the identity of mind knowing with the object known. There is a raying out of all orders of existence, an external emanation from the ineffable One. There is again a returning impulse, drawing all upwards and inwards towards the centre from whence all came.....

You ask, how can we know the Infinite? I answer, not by reason. It is the office of reason to distinguish and define. The Infinite, therefore, cannot be ranked among its objects. You can only apprehend the Infinite by a faculty superior to reason, by entering into a state in which you are your finite self no longer – in which the Divine Essence is communicated to you. This is Ecstasy. It is the liberation of your mind from its finite consciousness. Like only can apprehend like; when you thus cease to be finite, you become one with the Infinite. In the reduction of your soul to its simplest self, its divine essence, you realise this Union – this Identity.

But this sublime condition is not of permanent duration. It is only now and then that we can enjoy this elevation ... above the limits of the body and the world..... All that tends to purify and elevate the mind will assist you in this attainment and facilitate the approach and the recurrence of these happy intervals. There are, then, different roads by which this end may be reached. The love of beauty which exalts the poet; the devotion to the One and that ascent of science which makes the ambition of the philosopher; and that love in those prayers by which some devout and ardent soul tends in its moral purity towards perfection. These are the great highways conducting to that height above the actual and the particular, where we stand in the immediate presence of the Infinite, who shines out as from the deeps of the soul"

(Extract of a letter from Plotinus to Flaccus. R.A. Vaughan *Hours with the Mystics*, New York, Charles Scribner & Sons, 1903, Vol. I. pp 78-81.)

Metatime

Just as for Louis MacNeice there was for only for a little while 'neither up nor down', Plotinus emphasises that we can be in this state only fleetingly. The time difference between the life of the above two writers illustrates that there is nothing new under the sun and that great minds have been there already a long time ago. Plotinus anticipates three major themes of this book. One is that a faculty other than reason, whose object is to 'distinguish and define', must take us to the absolute. Another is that to attain the absolute the individual self must be left behind. Thirdly and superficially surprising, Plotinus's 'great highways' are very similar to the three paths, based on Indian philosophy and yoga, around which this chapter has been constructed.

The way to enter metatime is by "changing the conditions of perception and in this way approach to the real essence of things" (Ouspensky: 1981, 13). This is achieved in a 'metatime' which, although it cannot be 'known' conceptually, has an effect in our lives in a number of ways. How this manifests itself will vary according to our nature and talents.

A natural scientist may be able to use his understanding to penetrate the secrets of our observable world; the artist may be able to represent its essence to our imagination or the contemplative to become one with all. But all of us can experience the transcendent bliss of metatime in periods of intense intellectual, emotional or physical activity – or indeed of inactivity!

Chapter Eight

Conclusion

"Sudden in a shaft of sunlight
Even while the dust moves
There arises the hidden laughter
Of children in the foliage
Quick now, here, now, always –
Ridiculous the waste sad time
Stretching before and after."

Burnt Norton T.S. Eliot

"Row, row, row your boat gently down the stream
Merrily, merrily, merrily, Life is but a dream"

Anon

'Verily I say unto you, Except ye be converted, and become as little children, ye shall not enter into the kingdom of heaven.'

Gospel according to St. Matthew Chapter 18, v3

Toddlers live an almost eternal now with the past and the future as only dim notions. It is only in adolescence that the whole cycle of life and death becomes less fuzzy. Fully mature we can go on developing, with accompanying evolutions in our awareness of time. In J.B. Priestley's book *Man and Time*, he writes about his heightened sense of time fighting in the horrors of the Western front during the First World War and almost in the same breath refers to the innocent timelessness of children playing on the beach. The sad truth is that in everyday life as adults most of us can only recapture that magic in extremis.

Conclusion

Time as a quantity, as clock time, is of great benefit in the day-to-day business of life. On the other hand, it is of no use and can be positively injurious if our whole life is run by it. Time becomes viewed as a commodity, which can be bought and sold, or a currency as in 'time is money'. For example, in life in quantitative terms longevity becomes a goal in its own right regardless of its quality. Similarly efficiency of production, or transport is seen in terms of the amount of time 'saved'. Some of the best things in life are only so because they are slow or short-lived.

In education the current emphasis on preparing children for 'adult' life by cramming them full of knowledge useful from a work perspective rather than letting their natural curiosity flower is also harmful. Without that curiosity learning and then work become a chore. Stunted growth results from forced feeding.

So-called 'primitive' societies are often regarded as 'childish' because they nurture and maintain this innocence. For example, Steve Charleston, a Native American professor of theology, discusses this specifically with reference to the tyranny of time in Western culture, calling time an invention of the human mind for describing change and motion. He connects it with the destruction of family relationships, where we put aside a little bit of 'quality time' for children between the pressures of our other commitments as if "human relationships can be made warm in the microwave of quick encounters" (Charleston: 1989, 27-32)

A good part of a child's wonder is concerned with ideas related to time. Roger Penrose, the distinguished mathematician and writer, at the end of his book "concerning computers, minds and the laws of physics", writes: "CONCLUSION: A CHILD'S VIEW. In this book I have presented many arguments intending to show the untenability of the viewpoint – apparently rather prevalent in current philosophising – that our thinking is basically the same as the action of some very complicated computer....the explicit assumption is made that mere enaction of an algorithm can evoke *conscious awareness* ....some of the arguments that I have given in these chapters may seem tortuous and complicated...yet beneath all this technicality is the feeling that it is indeed 'obvious' that the *conscious* mind cannot work like a computer, even though much of what is involved in mental activity might do so.

This is the kind of obviousness that a child can see – though that child may, in later life, become browbeaten into believing that the obvious problems are 'non-problems', to be argued into non-existence by careful reasoning and clever choices of definition. Children sometimes see things clearly that are indeed obscured in later life. We often forget the wonder we felt as children when the care of the activities of the 'real world' have begun to settle upon our shoulders. Children are not afraid to pose basic questions that may embarrass us as adults to ask. What happens to our thoughts after we die, where was I was born.... when did it start ...why are we here; why is there a universe at all?

I remember, myself being troubled by many such puzzles as a child....perhaps only the present instance 'exists 'for me. Perhaps the 'me' of tomorrow, or of yesterday, is really quite a different person with an independent consciousness...." (Penrose: 1989, 578-581)

These are questions related to us internally as much as much as externally when we socialise and mix with people other than our own selves. Without any inner imagination a generation may emerge, who in fact could become to resemble more and more the inanimate devices that now occupy so much of our time and leisure – mobile phones, computers, screens for watching shows on television or other people on CCTV, other screens for endless visual images captured on our personal cameras and then posted on to social media and so preserved for all time, endless gadgets for cooking and cleaning and devices for moving around over the ground, in the air and on water, music from Ipods In the end all that human beings will do (or rather not do) is to passively react to input. Never ask why or what for – don't actually do anything, think anything or feel anything deeply or intensely.

But the external world can be put in its place as just the external reflection of our inner riches which is the stage on which the show of life takes place. Machines and words useful as means but it is not as ends. They are just tools.

Any theory of everything must be much broader than that aimed at by scientists. It must include the intangibles, the inexpressible and the transcendent. It must be idealist not realist and its findings should be truths not completely expressible in language, because they precede language and render it possible.

Let objective theories help us to improve our material conditions but let them not be conflated with what is truly important:

"The much needed broadening of the mind by science has only replaced medieval one-sidedness – namely the age-old unconsciousness which once predominated and has gradually become defunct – by a new one-sidedness, the overvaluation of 'scientifically' attested views. These each and all relate to objects in a chronically one-sided way, so that nowadays the backwardness of psychic development in general and of self-knowledge in particular has become one of the most pressing contemporary problems. As a result of the prevailing one-sidedness, and in spite of an appalling optical demonstration of an unconscious that has been alienated from the conscious, there are still vast numbers of people who are the blind and helpless victims of these conflicts" (Jung 2001: para 426).

Jung adds to the above quotation that the problem exists because people "apply their scientific scrupulosity only to external objects, never to their own psychic condition". This is Jung's 'faith' that the only answer must lie in more science. No! No! No! Pascal summed up the case against succinctly: "Le coeur a ses raisons que la raison ne connait point." Any scientific advances need to be balanced by emotional developments. Indeed, not just emotional but sensual, 'haptic' and spiritual. If we don't start paying attention to this we shall remain in our present unbalanced condition of being 'hollow men', with our internal development failing to keep abreast of our external development with its growing mastery of our physical makeup and of the world around us. A paradigm of time based exclusively around 'natural science' is a positive hindrance to any inner growth.

Bibliography

Anonymous (1982) *The Cloud of Unknowing*, Hodgson, Phyllis Ed. : Exeter: The Catholic Records Press (original late fourteenth century)

Appleyard, Brian (1994) *Understanding the Present*, New York: Anchor

Appleyard, Brian (2002) 'Brief encounters: Brian Appleyard's life-changing meeting
g with Stephen Hawking' *Sunday Times*, 27 January

Bachelard, Gaston (1935) *L'intuition de l'instant. Etude sur la Siloë de Gaston Roupnel*, Paris: Stock

Barbour Julian (1999) *The End of Time,* Weidenfeld & Nicholson: London

Bergson, Henri (1913) *Introduction to Metaphysics*, London: Macmillan

Bergson, Henri (1910) *Time and Free Will: An Essay on the Immediate Data of Consciousness,* London: Macmillan (French original, *Essai sur les données immediates de la conscience* 1889)

Bergson, Henri (1970*) Oeuvres, Edition du Centenaire,* Paris : Presses Universitaires de France

Bhagavad-Gita (1954) translated by Swami Prabhavananda and Christopher Isherwood, New York: Mentor Books

Blofeld, John (1970) *The Way of Power, A Practical Guide to the Tantric Mysticism of Tibet*, London: George, Allen & Unwin Ltd

Boethius (1969) *Consolation of Philosophy*, Harmondsworth: Penguin

Borges, Jorge Luis (1970) *Labyrinths: A New Refutation of Time*, Harmondsworth: Penguin

Bucke, Richard Maurice (1901) *Cosmic Consciousness: A Study of the Evolution of the Human Mind*, Bedford, Massachusetts: Applewood Books (classic reprint)

Capra, Fritjof (1976) *The Tao of Physics*, London: Fontana Paperback

Capra, Fritjof (1983) *The Turning Point*, London: Flamingo

Carter, Rita (1998) *Mapping the Mind*, London: Weidenfield & Nicolson

Cartwright, Nancy (1999) *The Dappled World*, Cambridge University Press

Chatwin, Bruce (1987) *The Songlines*, London: Jonathan Cape

Charleston, S (1989) 'The Tyranny of Time', *Lutheran Women Today* 2 (7), 27-32

Collin, Rodney (1954) *The Theory of Celestial influence*, London: Vincent Stuart

Coomaraswamy (2001) *Time and Eternity*, New Delhi: Munshiram Manoharial Ltd.

Coveney, Peter (1991) 'Chaos, Entropy and the Arrow of Time', *New Scientist Guide to Chaos*

Coveney, Peter & Highfield, Roger (1990) *The Arrow of Time*, London: W.H.Allen

Damasio Antonio (2000) *The Feeling of What Happens,* London: Vintage,

Damasio, Antonio (2006) 'Remembering When', *Scientific American, Special Issue, A Matter of Time*, Vol 287, Issue 3

Davies, Paul (1980) *Other Worlds*, Harmondsworth: Penguin

Davies, Paul (1995) *About Time: Einstein's Unfinished Revolution*, Harmondsworth: Penguin

Davies, Paul (1999) *The Mind of God*, Harmondsworth: Penguin

Davies, Paul (2002) 'That Mysterious Flow', *Scientific*

Bibliography

American, Special Issue, A Matter of Time, Vol 287, Issue 3

Dawkins, Richard (1976) *The Selfish Gene,* Oxford University Press

Dawkins, Richard (1998) *Unweaving the Rainbow: Science, Delusion and The Appetite for Wonder,* London: Allen Lane

Deleuze, Gilles (1963) *La philosophie critique de Kant,* Presses Universitaires de France

Deleuze, Gilles (1989) *Cinema 2: The Time Image,* (French original, *Cinema II. L'image-temps,* 1985: Paris)

Dennett, Daniel (1993) *Consciousness Explained,* Harmondsworth: Penguin

Dennett, Daniel (2002) 'The Grand Illusion. How could I be wrong? How wrong could I be?' *Journal of Consciousness Studies,* January

Dunne, J.W. (1934) *The Serial Universe,* London: Faber and Faber

Eccles, J.C (1953) *The Neurophysiological Basis of Mind,* OUP

Eco, Umberto (1989) *Foucault's Pendulum,* London: Secker and Warburg

Gibson, James (1966) *The Senses considered as Perceptual Systems,* Houghton Mifflin & Co

Gleick (1987) *Chaos: Making a New Science,* New York ; Viking Penguin

Goldschmidt, Victor (1969) *Le Système Stoicien et L□Idée de Temps,* Paris: Vrin

Goswami, Amit (1993) *The Self-Aware Universe,* Penguin

Goswami, Amit (1998) 'From Deep to Deepest Ecology', *Bulletin of Science within Consciousness,* Volume 3.no 2, Fall 1998

Greene, Brian (1999) *The Elegant Universe,* London: Jonathan Cape

Griffiths, Jay (1999) *Pip-Pip,* London: Flamingo

Haldane, J.B.S (1940) *Possible Worlds,* London: Evergreen Books

Hall, Edward T (1984) *The Dance of Life,* New York: Anchor Books

Hamburger, Michael (1972) *The Truth of Poetry,* Harmondsworth: Penguin

Hampshire, Stuart (1959) *Thought and Action,* London: Chatto and Windus

Happold, F.C (1963) *Mysticism: A Study and an Anthology,* Harmondsworth: Penguin

Hawking, Stephen (1988) *A Brief History of Time,* London: Bantam Books

Hawking, Stephen (2001) 'Back to the future with a big bang', *Daily Telegraph,* Oct 24

Hawking, Stephen (2001) *The Universe in a Nutshell,* Random House

Heidegger, Martin (1962) *Being and Time,* Oxford: Blackwell (German Original, 1927)

Heidegger, Martin 1962 *The Concept of Time,* Oxford: Blackwell (German Original, 1924)

Hume, David (1894) *An Inquiry Concerning Human Understanding,* (Sections 7 & 8) ed. Selby-Bigg, OUP

Hume, David (1967) A *Treatise on Human Nature,* ed. Selby-Bigg, OUP

Huxley, Aldous (1959) *The Doors of Perception and Heaven and Hell,* Harmondsworth: Penguin one volume edition

Huxley, Aldous (1958) *The Perennial Philosophy,* London & Glasgow: Collins

James, William (1950) *The Principles of Psychology. Volume One,* New York: Dover Publications (Original, 1890)

Bibliography

James, William (1917) *The Varieties of Religious Experience*, London: Longmans

Jung, Carl Gustav (1960) *The Collected Works of C.G. Jung*, eds., Herbert Read, Michael Fordham & Gerard Adler, London: Routledge & Kegan Paul

Jung Carl Gustav (2001) *On the Nature of the Psyche*, London: Routledge Classics

Kant, Immanuel (1972) *The Moral Law: Kant's Groundwork of the Metaphysic of Morals*, translated by H.J. Paton, London: Hutchinson (German original, 1785)

Kant, Immanuel (1934) *A Critique of Pure Reason*, translated by J.M.D.Meiklejohn, London: Dent (German original, 1781)

Kern, Stephen (1983) *The Culture of Time and Space 1880-1918*, Harvard University Press

Kleist, Heinrich von (1994) 'On the Marionette Theatre' in *Essays on Dolls, Heinrich von Kleist Charles Baudelaire, Rainer Maria* Rilke1-12 , translated by Idris Parry and Paul Kegan, Harmondsworth: Penguin (German original, 1810)

Koestler, Arthur (1959) *The Sleepwalkers*, London: Hutchinson

Koestler, Arthur (1976) *The Ghost in the Machine*, Revised Edition, London: Hutchinson,

Langer, Susanne (1953) *Feeling and Form*, London: Routledge, Kegan & Paul

Le Poidevin, Robin (2000) 'The Experience and Perception of Time' in *The Stanford Encyclopaedia of Philosophy*, Edward N. Zalta (ed.), URL

Lovat, T.J. (2003) 'The distinctive contribution of proportionism to scientific assessment in a post-scientific age', *Theology & McCauley* Issue 3, McCauley College, Queensland, Australia.

Mascaro, Juan (1965) *The Upanishads*, Harmondsworth: Penguin

Merleau-Ponty, Maurice (1964) 'Eye and Mind', translated by Carleton Dallery, in *Maurice Merleau-Ponty The Primacy of Perception,* ed. J.M.Eddie, North West University Press: Evanston

Midgley, Mary (2001) *Science and Poetry,* London: Routledge.

Nadeau, Robert & Kafatos, Menas (2001) *The Non-Local Universe. The New Physics and Matters of the Mind,* OUP

Nagel, Thomas (1974) 'What is it like to be a Bat', *Philosophical Review,* 435-50

Newton, Isaac (1999) *The Principia,* University of California Press (Latin original, 1687)

Ouspensky, P.D. (1934) A *New Model of the Universe,* London: Routledge & Kegan Paul

Ouspensky, P.D. (1950) *In Search of the Miraculous,* London: Routledge & Kegan Paul

Ouspensky, P.D. (1981) *Tertium Organum,* London: Routledge & Kegan Paul

Peat, F. David (1991) 'Non-Locality in Nature and Cognition' *Nature and Cognition,* ed. M. Karvallo Kulwer, Academic Publishers

Penrose, Roger (1989) *The Emperor's New Mind,* OUP Paperback

Penrose, Roger (1994) *Shadows of the Mind, A Search for the Missing Science of Consciousness,* OUP

Penrose, Roger (2004) *The Road to Reality: A Complete Guide to the Laws of the Universe,* London: Jonathan Cape

Picqot, Thierry (2003) *The Art of the Siesta,* translated by Ken Hollings, London: Marion Boyars

Plato (1997) *Complete Works,* eds. John M. Cooper, D. S. Hutchinson, Hackett Pub

Plotinus (1991) *The Enneads,* translated by Stephen

Bibliography

McKenna, Harmondsworth: Penguin

Popper, Karl (1963) *Conjectures and Refutations,* London: Routledge

Priestley, J.B. (1964) *Man and Time*, London: Aldus Books

Price, Huw (1996) *Time's Arrow and Archimedes' Point*, OUP

Rae, A (1994) *Quantum Physics, Illusion or Reality*, Canto edition, CUP

Rilke, Rainer Maria (1949) *The Notebooks of Malte Laurids Brigge*, translated by M.D. Herter New York: W.W. Norton & Co,

Ruckner, Rudy (1986) *The Fourth Dimension and How to Get There*, Harmondsworth: Penguin

Russell, Bertrand (1963) *Mysticism and Logic*, Unwin Books: London

Sandbothe M.von (2001) *The Temporalization of Time*, Oxford: Rowman & Littlefield

Sebald, W.G (2001) *Austerlitz*, Harmondsworth: Penguin

Smart, J.J.C (1964) *Problems of Space and Time*, London: Macmillan

Spengler, Oswald (1932) Decline *of the West*, translated by Charles Francis Atkinson, London: George Allen & Unwin (German original, 1917)

Steiner, George (1997) *Errata*, London: Weidenfeld & Nicolson

Suzuki, Daisetz (1934*) Manual of Zen Buddhism*, London: Hutchinson,

Thompson (1990) *Vedic Cosmography and Astronomy*, Los Angeles The Bhaktivedanta Book Trust

Trimingham, Jolyon (2005) *Poetry and Poetical Language: Love, Death, Dust and Time*, unpublished essay

Trimingham, Melissa (2001) *The Practical Application of principles underlying the work of*

Oskar Schlemmer at the Dessau Bauhaus 1926-29, Doctoral Thesis University of   Leeds

Whitehead, Alfred (1938) *Modes of Thought*, Cambridge University Press

Whitrow, G.J. (1961) *The Natural Philosophy of Time*, London: Thomas Nelson

Whitrow, G.J. (1975) *The Nature of Time*, Harmondsworth: Penguin

Whitrow, G.J. (1989) *Time in History,* OUP Paperback

Wolfram, Stephen (2001) *A New Kind of Science*, Champaign: Wolfram Media Inc

Zohar, Danah (1990) *The Quantum Self*, London: Bloomsbury Publishing

Zuckerkandl, Victor (1956) Sound *and Symbol*, Princeton UP)

Index of Names

W

Welles, Orson 90, 92
Whitehead, Alfred 102
Whitrow, G.J. 46, 101
Wolfram, Stephen 9, 192
Wordsworth, William 165, 166

Y

Yeats, W.B. 39

Z

Zeno of Elea 98, 99
Zuckerkandl, Victor 77, 78, 79,
          80, 147, 192

Author

Jolyon Trimingham

Jolyon Trimingham's formative years were dominated by his wildly eccentric Canadian father, a prolific inventor. Removed from his school in London for a year's round-the-world trip with him when he was thirteen, Jolyon did not break free of his father's unconventional and nomadic influence until his early twenties when he left his father's workshop and went to live in Paris.

"I developed a special interest in three writers, Immanuel Kant, the German philsopher of history, Oswald Spengler, and the twentieth century mystic Russian writer P.D. Ouspensky. The seeds of *Understanding Time* were planted in the reading of their works. When discussing the nature of time they seemed to be talking about different subjects, but the idea dawned upon me that there must be a kind of paradigm shift which could link Spengler's chronological time, Ouspensky's mystical time and Kant's time centred round mathematics and natural philosophy".

After this first period in Paris Jolyon went on to study Politics, Philosophy and Economics at Oxford University as a mature student, and wrote his undergraduate thesis on Henri Bergson's *'Time and Free Will'*. Ten years later he returned to Paris to study philosophy at the Sorbonne.

Outside of his career as a technical writer, which has taken him all over Europe, Jolyon has found time to restore an open-air Regency swimming pool in Bath, launch a successful bid to save a hill top lake in Yorkshire from being drained, and stood as the Green Party parliamentary candidate for Dover in 2015.